WATER-PUMPING DEVICES

WATER-PUMPING DEVICES

A handbook for users and choosers

Peter Fraenkel

INTERMEDIATE TECHNOLOGY PUBLICATIONS by arrangement with
THE FOOD AND AGRICULTURE ORGANIZATION OF THE UNITED
NATIONS 1986

Intermediate Technology Publications
9 King Street, London WC2E 8HW, U.K.

© FAO 1986

This book is published by arrangement with the Food and Agriculture Organization of the United Nations (FAO), in whose *Irrigation and Drainage Paper* series it appears.

ISBN 0 946688 85 0

Printed in England

CONTENTS

Page

1. INTRODUCTION

   1.1 Scope and purpose of this paper 1
   1.2 The increasing importance of irrigation 1
   1.3 Irrigation and the "Energy Crisis" 2
   1.4 Small-scale irrigation and development 3
   1.5 The choice of water lifting technique 4

2. WATER LIFTING FOR IRRIGATION 5

   2.1 General principles of water lifting 5
        2.1.1 Definitions of Work, power, energy and efficiency . . 5
        2.1.2 Efficiency of components: the importance of matching 6
        2.1.3 Irrigation system losses 7
        2.1.4 Flow through channels and pipes 8
        2.1.5 Suction lift: the atmospheric limit 13
        2.1.6 Drawdown and seasonal variations of water level . . . 13
        2.1.7 Review of a complete lift irrigation system 15
        2.1.8 Practical power requirements 15

   2.2 Outline of principles of small-scale irrigation 18
        2.2.1 Irrigation water requirements 18
        2.2.2 Nett irrigation requirement 18
        2.2.3 Gross irrigation requirement 19
        2.2.4 Pumping requirement 20

3. REVIEW OF PUMPS AND WATER LIFTING DEVICES 21

   3.1 Principles for moving or lifting water 21
   3.2 Taxonomy of water lifts and pumps 21
   3.3 Reciprocating and cyclic direct lift devices 22
        3.3.1 Watering cans, buckets, scoops, bailers and the
               swing basket 22
        3.3.2 Suspended scoop, gutters, dhones and the counterpoise-
               -lift or shadoof 22
        3.3.3 Bucket hoists, windlasses, mohtes and water skips . . 22

   3.4 Rotary direct lift devices 23
        3.4.1 Bucket elevators, Persian wheels and norias 23
        3.4.2 Improved Persian wheels, (zawaffa or jhallar) 24
        3.4.3 Scoop-wheels; sakia, tympanum or tablia 24

3.5	Reciprocating displacement pumps		26
	3.5.1	Piston or bucket pumps: basic principles	26
	3.5.2	Double-acting piston pumps and plunger pumps	29
	3.5.3	Pistons and valves	31
	3.5.4	Reciprocating pumps and pipelines	32
	3.5.5	Reciprocating borehole pumps	35
	3.5.6	Hydraulically activated borehole pumps	36
	3.5.7	Diaphragm pumps	37
	3.5.8	Semi-rotary pumps	38
	3.5.9	Gas displacement pumps	38
3.6	Rotary positive displacement pumps		39
	3.6.1	Flexible vane pumps	40
	3.6.2	Progressive cavity (Mono) pumps	40
	3.6.3	Archimedean screw and open screw pumps	41
	3.6.4	Coil and spiral pumps	42
	3.6.5	Paddle wheels, treadmills and flashwheels	43
	3.6.6	Water-ladders and Dragon Spine pumps	44
	3.6.7	Chain and washer or Paternoster pumps	45
3.7	Reciprocating inertia (joggle) pumps		46
	3.7.1	Flap valve pump	47
	3.7.2	Resonant joggle pump	47
3.8	Rotodynamic pumps		48
	3.8.1	Rotodynamic pumps: basic principles	48
	3.8.2	Volute, turbine and regenerative centrifugal pumps	48
	3.8.3	Rotodynamic pump characteristics and impeller types	50
	3.8.4	Axial flow (propeller) pumps	51
	3.8.5	Mixed flow pumps	52
	3.8.6	Centrifugal pumps	53
	3.8.7	Multi-stage and borehole rotodynamic pumps	55
	3.8.8	Self-priming rotodynamic pumps	56
	3.8.9	Self-priming jet pumps	56
3.9	Air-lift pumps		58
3.10	Impulse (water hammer) devices		59
3.11	Gravity devices		59
	3.11.1 Syphons		59
	3.11.2 Qanats and foggara		60
3.12	Materials for water lifting devices		61
3.13	Summary review of water lifting devices		65
4.	POWER FOR PUMPING		66
	4.1	Prime-movers as part of a pumping system	66
		4.1.1 Importance of cost-effectiveness	66
		4.1.2 Transmission systems	68
		4.1.3 Fuels and energy storage	71

4.2	Human power	71
	4.2.1 Human beings as power sources	71
	4.2.2 Traditional water lifting devices	74
	4.2.3 Handpumps	75
	4.2.4 Handpump maintenance	77
4.3	Animal power	78
	4.3.1 Power capabilities of various species	79
	4.3.2 Food requirements	80
	4.3.3 Coupling animals to water-lifting systems	81
4.4	Internal combustion engines	83
	4.4.1 Different types of i.c. engine	83
	4.4.2 Efficiency of engine powered pumping systems	88
4.5	External combustion engines	91
	4.5.1 Steam engines	92
	4.5.2 Stirling engines	93
4.6	Electrical power	94
	4.6.1 Sources and types of electricity	95
	4.6.2 AC mains power	97
	4.6.3 Electric motors	97
	4.6.4 Electrical safety	99
4.7	Wind power	99
	4.7.1 Background and State-of-the-Art	99
	4.7.2 Principles of Wind Energy Conversion	102
	4.7.3 The Wind Resource	109
	4.7.4 Windpump Performance Estimation	115
4.8	Solar power	118
	4.8.1 Background and State-of-the-Art	118
	4.8.2 Principles of solar energy conversion	120
	4.8.3 The solar energy resource	127
	4.8.4 Performance estimation	127
4.9	Hydro power	130
	4.9.1 Background and State-of-the-Art	130
	4.9.2 Use of turbines for water lifting	133
	4.9.3 The hydraulic ram pump (or hydram)	135
	4.9.4 Water wheels and norias	138
	4.9.5 Novel water powered devices	140
4.10	Biomass and coal (the non-petroleum fuels)	142
	4.10.1 The availability and distribution of fuels	142
	4.10.2 The use of solid fuels	145
	4.10.3 The use of liquid biomass fuels	149
	4.10.4 Gas from biomass: Biogas	151

| 5. | THE CHOICE OF PUMPING SYSTEMS | 156 |

	5.1	Financial and economic considerations	156
		5.1.1 Criteria for cost comparison	156
		5.1.2 Calculation of costs and benefits	157
		5.1.3 Relative economics of different options	163
	5.2	Practical considerations	166
		5.2.1 Status or availability of the technology	166
		5.2.2 Capital cost versus recurrent costs	167
		5.2.3 Operational convenience	168
		5.2.4 Skill requirements for installation, operation and maintenance	168
		5.2.5 Durability, reliability and useful life	169
		5.2.6 Potential for local manufacture	169
	5.3	Conclusion	170

REFERENCES   171

ACKNOWLEDGEMENTS

The publishers would like to thank the following especially for the use of material from which illustrations have been drawn:

BYS, Napel; Grundfos; Khan, H.R. and the Institute of Civil Engineers, London; Lysen, E. and CWD, The Netherlands; National Aeronautical Lab., Bangalore; Smithsonian Institute; UN Economic and Social Commission for Asia and the Pacific.

1. INTRODUCTION

1.1 SCOPE AND PURPOSE OF THIS PAPER

This paper in effect replaces the excellent, but now long out of print booklet, entitled "Water Lifting Devices for Irrigation", by Aldert Molenaar, published by FAO as long ago as 1956 [1]. Since that time, little more than one generation ago, the human population has almost doubled. In the same short period, over twice as much petroleum, our main source of energy, has been consumed as in the whole of history prior to 1956. But there has also been a much wider awareness of the constraints which must force changes in technology.

The primary purpose of this paper is to provide a basis for comparing and choosing between all present and (near) future options for lifting irrigation water on small and medium sized land-holdings (generally in the range 0.25 ha to say 25 ha). Small land-holdings in this size range are most numerous in many of the developing countries, and extension of the use of irrigation in this small farming sector could bring huge benefits in increased food production and improved economic well-being. It is also hoped that this paper will be useful to those seeking techniques for lifting water for purposes other than irrigation.

1.2 THE INCREASING IMPORTANCE OF IRRIGATION

Water has always been a primary human need; probably the first consideration for any community has always been the need for ready access to it. Irrigation water more specifically can offer the following important benefits:

i. increases land area brought under cultivation

ii. improves crop yield over rain-fed agriculture three or four-fold

iii. allows greater cropping intensity

iv. produces improved economic security for the farmer

v. reduced drought risk, which in turn allows:

    - use of high yield seeds
    - increased use of fertilizer, pesticides and mechanization
    - control of timing for delivery to market
    - control of timing for labour demand

vi. allows introduction of more valuable crops

Feeding the rapidly growing human race is an increasingly vital problem. There is no readily identifiable yield-increasing technology other than the improved seed-water-fertilizer approach. It is expected that in the next two decades about three quarters of all the increases in the output of basic staples will have to come from yield increases, even though during the past decade yield increases have only succeeded in supplying half the increase in output [2]. This is because there is less and less fertile but as yet uncultivated land available in the more densely populated regions. Irrigation of crops is a primary route to bringing more land under cultivation and to increasing yields from existing farm land. Irrigation will therefore be increasingly important in the future both to increase the yield from already cultivated land and also to permit the cultivation of what is today marginal or unusable land.

Table 1 indicates the irrigated regions of the world; (adapted from [3]), and the principal developing countries where irrigation is currently practised. The majority of the land brought under irrigation since 1972 is mainly in countries where irrigation is already generally practised. Not many countries have significant areas of irrigated land and the two most populous countries, China and India, have about half of the entire world's irrigated land area within their borders. These two large and crowded countries will have to increase their irrigated land still further to improve their food production, while other countries facing similar population pressures on the land will have to do tomorrow what India and China do today.

1.3   IRRIGATION AND THE "ENERGY CRISIS"

Water and good land can often be found in juxtaposition, but it is the provision of the necessary power for pumping which is so often the primary constraint. Human muscle power or domestic animals have been used since antiquity, and still are being used in many parts of the world, to lift and distribute water, but as will be explained later, these techniques are often extremely costly in real terms due to the low productivity that is achieved. Therefore, mechanized lift irrigation techniques are becoming increasingly important to meet the enormous predictable future demand.

The area of irrigated land in the world has been estimated to have increased by about 70% in the period 1952 to 1972 [3] and much of this expansion will have been through the increasing use of engine and mains-electrified pumps during that period of decreasing fuel and electricity prices (in real terms). However, since then the price of petroleum, and hence of electricity, has tended to rise, and this has reduced the margin to be gained by farmers from irrigation, since food prices have generally been prevented from rising in line with energy costs. Some governments attempt to mitigate this situation by subsidizing oil and rural electricity for use in agriculture, but many of these governments are the very ones that can least afford such a policy which exacerbates balance of payments deficits by encouraging the use of oil.

Despite present short-term fluctuations in oil prices, conventional oil-based engine-driven power sources and mains electricity are expected to continue to increase in the longer term. There are also major problems associated with maintenance of this kind of machinery.

There is therefore a considerable incentive in most of the poorer developing countries to discourage the use of oil, even though there is an equally strong incentive to encourage the increase of agricultural production,

Table 1 IRRIGATED AREAS OF THE WORLD (1972)

REGION & principal irrigation countries	IRRIGATED AREA million hectares (Mha)	% of total
1. SOUTH & S.E. ASIA	132	66
China	74	
India	33	
Pakistan	12	
Indonesia	4	
Taiwan	2	
Thailand	2	
2. NORTH AMERICA	17	9
3. EUROPE	13	7
4. MIDDLE EAST	11	5
Iraq	4	
Iran	3	
Turkey	2	
5. USSR	10	5
6. AFRICA	7	3
Egypt	3	
Sudan	1	
7. CARIBBEAN & CENTRAL AMERICA	5	2
Mexico	4	
8. SOUTH AMERICA	4.5	2
Argentina	1.2	
Chile	1.3	
9. AUSTRALASIA	1.4	1
WORLD TOTAL	201.9	100

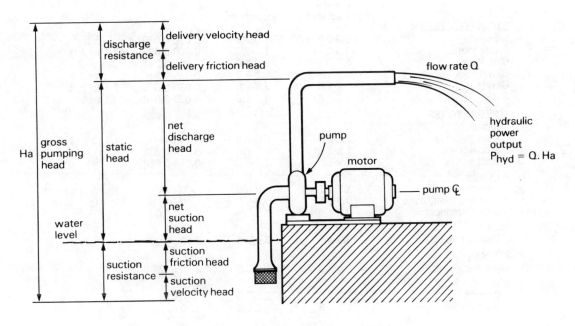

Fig. 1   Typical pump installation

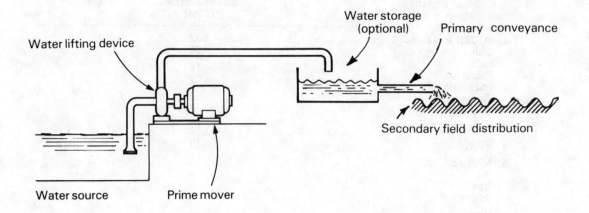

Fig. 2   Key components of an irrigation system

Table 2   SUGGESTED MAXIMUM FLOW VELOCITIES, COEFFICIENTS OF ROUGHNESS AND SIDE SLOPES, FOR LINED AND UNLINED DITCHES AND FLUMES

Type of surface	Maximum flow velocities		Coefficients of roughness (n)	Side slopes or shape
	Metres per second	Feet per second		
UNLINED DITCHES				
Sand	0.3-0.7	1.0-2.5	0.030-0.040	3:1
Sandy loam	0.5-0.7	1.7-2.5	0.030-0.035	2:1 to 2½:1
Clay loam	0.6-0.9	2.0-3.0	0.030	1½:1 to 2:1
Clays	0.9-1.5	3.0-5.0	0.025-0.030	1:1 to 2:1
Gravel	0.9-1.5	3.0-5.0	0.030-0.035	1:1 to 1½:1
Rock	1.2-1.8	4.0-6.0	0.030-0.040	¼:1 to 1:1
LINED DITCHES				
Concrete				
Cast-in-place	1.5-2.5	5.0-7.5	0.014	1:1 to 1½:1
Precast	1.5-2.0	5.0-7.0	0.018-0.022	1½:1
Bricks	1.2-1.8	4.0-6.0	0.018-0.022	1½:1
Asphalt				
Concrete	1.2-1.8	4.0-6.0	0.015	1:1 to 1½:1
Exposed membrane	0.9-1.5	3.0-5.0	0.015	1½:1 to 1:1
Buried membrane	0.7-1.0	2.5-3.5	0.025-0.030	2:1
Plastic				
Buried membrane	0.6-0.9	2.0-3.0	0.025-0.030	2½:1
FLUMES				
Concrete	1.5-2.0	5.0-7.0	0.0125	
Metal				
Smooth	1.5-2.0	5.0-7.0	0.015	
Corrugated	1.2-1.8	4.0-6.0	0.021	
Wood	0.9-1.5	3.0-5.0	0.014	

which so often demands pumpeded irrigation. As a result, there is an increasing need to find methods for energizing irrigation pumps that are independent of imported oil or centralized electricity.

1.4   SMALL-SCALE IRRIGATION AND DEVELOPMENT

Intensive irrigation of small-holdings is likely to become increasingly important and widely used during the next few decades, particularly in the developing countries. This is because the majority of land-holdings, particularly in Asia and Africa are quite small, under 2 ha [4]. Even in South America, where the maximum percentage of farmed land consists of very large land-holdings, the most numerous type of land-holding is under 5 ha.

Studies have shown that small land-holdings are often more productive, in terms of yield per hectare, than larger units. An Indian farm management study [5], indicated that small family run land-holdings are consistently more productive than larger units, although they are more demanding in terms of labour inputs. A similar survey in Brazil [5], also showed better land utilization on small land-holdings; however this was achieved by applying between 5 and 22 times as much labour per hectare compared with large farms

Small land-holdings also generally achieve better energy ratios than large ones; i.e. the ratio of energy available in the crop produced, to the energy required to produce it. Energy ratios for tropical subsistence and semi-subsistence agriculture are in the range 10 to 60 (i.e. the food product has 10 to 60 times as much energy calorific value as the energy input to grow it) [6]. Mechanised large scale commercial agriculture, which usually, but not necessarily produces a better financial return, generally has energy ratios in the range from about 4 to less than 1. Therefore, in a situation where commercial fuels will get both scarcer and more expensive, there is more scope for increasing food production through improving the productivity of small labour-intensive land-holdings which have the potential capability to produce most food from a given investment in land and energy.

Small-scale irrigation has been shown to offer positive results in alleviating poverty. For example, the introduction of irrigation can double the labour requirements per hectare of land [5], and raise the incomes thereby not only of the farmers but also of landless labourers. The same reference gives examples from actual surveys of the average percentage increase in income for farmers who practised irrigation compared with those who did not; examples of increases obtained were 469% in Cameroon, 75% in South Korea, 90% in Malaysia, and 98% in Uttar Pradesh, India. In the Malaysian case, the increased income for landless labourers resulting from the introduction of irrigation averaged 127%.

Finally, there is probably more scope for significantly increasing yields in the small farm sector through irrigation than with large farms. For example, the average rice yield in the poorer South and South East Asian countries is typically 2 t/ha, while in Japan, with sophisticated small-scale irrigation and land management, 6 t/ha is commonly achieved [7]. The Asian Development Bank has reported that a doubling of rice production per hectare should be possible in the region within 15 years [7]. Obviously irrigation is not the only factor necessary to achieve such improvements, but it is perhaps one of the primary needs.

1.5 THE CHOICE OF WATER LIFTING TECHNIQUE

There are many different types of human and animal powered water lift, some of which are better than others for different purposes. While the power source or prime-mover so often attracts most interest, the correct selection of water conveyance and field distribution system can often have a greater influence on the effectiveness (technically and economically) of any irrigation system than differences between pumping power sources. In fact the use of a well-optimized and efficient water distribution system is vital when considering certain renewable energy systems where the cost is closely related to the power rating, and therefore a minimum power system needs to be selected.

Before looking for radical new water lifting techniques, there is also much scope for improving traditional and conventional pumping and water distribution methods; for example, petroleum-fuelled engines are commonly badly matched to both the pump and the piping system used for water distribution, which can waste a considerable proportion of fuel used.

The wide range of options for providing power for pumping water include some traditional technologies, such as windmills, and some entirely new technologies owing their origins to very recent developments, such as solar photovoltaic powered pumps. There are also technologies which have been widely and successfully used in just one area but which remain unknown and unused elsewhere with similar physical conditions; an example is the hydro-powered turbine pump, which has been used in tens of thousands solely in China. There are also some interesting new (and some not so new) options which are currently being experimented with, some of which may become available for general use in the near future; for example, steam pumps, Stirling engine pumps, and gasifiers for running internal combustion engines. All of these can produce pumping power from agricultural residues or other biomass resources, perhaps in future even from fuel crops, and may become more important as oil becomes scarcer and more expensive.

2. WATER LIFTING FOR IRRIGATION

2.1 GENERAL PRINCIPLES OF WATER LIFTING

2.1.1 Definitions of Work, Power, Energy and Efficiency

Energy is required, by definition, to do work; the rate at which it is used is defined as power (see Annex I for detailed definitions of units and their relationships). A specific amount of work can be done quickly using a lot of power, or slowly using less power, but in the end the identical amount of energy is required (ignoring "side-effects" like efficiencies).

The cost of pumping or lifting water, whether in cash or kind, is closely related to the rate at which power is used (i.e. the energy requirement in a given period). Since there is often confusion on the meaning of the words "power" and "energy", it is worth also mentioning that the energy requirement consists of the product of power and time; for example, a power of say, 5kW expended over a period of say, 6h (hours), represents an energy consumption of 30kWh (kilowatt-hours). The watt (W), and kilowatt (kW) are the recommended international units of power, but units such as horsepower (hp) and foot-pounds per second (ft.lb/s) are still in use in some places. The joule (J) is the internationally recommended unit of energy; however it is not well known and is a very small unit, being equivalent to only 1 Ws (watt-second). For practical purposes it is common to use MJ (megajoules or millions of joules), or in the world outside scientific laboratories, kWh (kilowatt-hours). 1kWh (which is one kilowatt for one hour or about the power of two horses being worked quite hard for one hour) is equal to 3.6MJ. Fuels of various kinds have their potency measured in energy terms; for example petroleum fuels such as kerosene or diesel oil have a gross energy value of about 36MJ/litre, which is almost exactly 10kWh/litre. Engines can only make effective use of a fraction of this energy, but the power of an engine will even so be related to the rate at which fuel (or energy) is consumed.

The hydraulic power required to lift or pump water is a function of both the apparent vertical height lifted and the flow rate at which water is lifted.

where $P_{hyd} = \rho g H_a Q$ is the hydraulic power
and ρ = density of water
g = acceleration due to gravity
H_a = vertical height
Q = flow rate

In other words, power needs are related <u>pro rata</u> to the head (height water is lifted) and the flow rate. In reality, the actual pumping head imposed on a pump, or "gross working head", will be somewhat greater than the actual vertical distance, or "static head", water has to be raised. Fig. 1 indicates a typical pump installation, and it can be seen that the gross pumping head, (which determines the actual power need), consists of the sum of the friction head, the velocity head and the actual static head (or lift) on both the suction side of the pump (in the case of a pump that sucks water) and on the delivery side.

The friction head consists of a resistance to flow caused by viscosity of the water, turbulence in the pump or pipes, etc. It can be a considerable source of inefficiency in badly implemented water distribution systems, as it is a function which is highly sensitive to flow rate, and particularly to pipe diameter, etc. This is discussed in more detail in Section 2.1.4.

The velocity head is the apparent resistance to flow caused by accelerating the water from rest to a given velocity through the system; any object or material with mass resists any attempt to change its state of motion so that a force is needed to accelerate it from rest to its travelling velocity. This force is "felt" by the pump or lifting device as extra resistance or head. Obviously, the higher the velocity at which water is propelled through the system, the greater the acceleration required and the greater the velocity head. The velocity head is proportional to the square of the velocity of the water. Therefore, if the water is pumped out of the system as a jet, with high velocity (such as is needed for sprinkler irrigation systems), then the velocity head can represent a sizeable proportion of the power need and hence of the running costs. But in most cases where water emerges from a pipe at low velocity, the velocity head is relatively small.

2.1.2 Efficiency of Components: the Importance of Matching

The general principle that:

   power   = (head x flowrate)

and   energy = (head x total weight of water lifted)

applies to any water lifting technique, whether it is a centrifugal pump, or a rope with a bucket on it. The actual power and energy needs are always greater then the hydraulic energy need, because losses inevitably occur when producing and transmitting power or energy due to friction. The smaller the friction losses, the higher the quality of a system. The quality of a system in terms of minimizing losses is defined as its "efficiency":

where efficiency $= \frac{\text{(hydraulic energy output)}}{\text{(actual energy input)}}$

using energy values in the equation gives the longer-term efficiency, while power values could be used to define the instantaneous efficiency.

A truly frictionless pumping system would in theory be 100% efficient; i.e. all the energy applied to it could reappear as hydraulic energy output. However, in the real world there are always friction losses associated with every mechanical and hydraulic process. Each component of a pumping system has an efficiency (or by implication, an energy loss) associated with it; the system efficiency or total efficiency is the product of multiplying together the efficiencies of all the components. For example, a small electrically driven centrifugal pump consists of an electric motor, (efficiency typically 85%), a mechanical transmission (efficiency if direct drive of say 98%), the pump itself (optimum efficiency say 70%) and the suction and delivery pipe system (say 80% efficient). The overall system efficiency will be the product of all these component efficiencies. In other words, the hydraulic power output, measured as (static head) x (flow) (since pipe losses have been considered as a pipe system efficiency) will in this case be 47%, derived as follows:

$$0.85 \times 0.98 \times 0.7 \times 0.8 = 0.47$$

The efficiency of a component is generally not constant. There is usually an operating condition under which the efficiency is maximized or the losses are minimized as a fraction of the energy throughput; for example a centrifugal pump always has a certain speed at a given flow rate and head at which its efficiency is a maximum. Similarly, a person or draft animal also has a natural speed of operation at which the losses are minimized and pumping is easiest in relation to output.

Therefore, to obtain a pumping system which has a high overall efficiency depends very much on combining a chain of components, such as a prime-mover, transmission, pump and pipes, so that at the planned operating flow rate and static head, the components are all operating close to their optimum efficiencies - i.e. they are "well matched". A most important point to consider is that it is common for irrigation systems to perform badly even when all the components considered individually are potentially efficient, simply because one or more of them sometimes are forced to operate well away from their optimum condition for a particular application due to being wrongly matched or sized in relation to the rest of the system.

2.1.3 Irrigation System Losses

The complete irrigation system consists not only of a water source and water lifting mechanism and its prime-mover and energy supply, but then there must also be a water conveyance system to carry the water directly to the field or plots in a controlled manner according to the crop water requirements. There may also be a field distribution system to spread the water efficiently within each field. In some cases there could be a water storage tank to allow finite quantities of water to be supplied by gravity without running the water lifting mechanism. Fig. 2 indicates the key components of any irrigation system, and also shows some examples of common options that fulfil the requirements and which may be used in a variety of combinations.

Most of the irrigation system components influence the hydraulic power requirements. For example, if pipes are used for distribution, even if they transfer water horizontally, pipe friction will create an additional resistance "felt" at the pump, which in effect will require extra power to overcome it. If open channels are used, extra power is still needed because although the water will flow freely by gravity down the channel, the input end of the channel needs to be high enough above the field to provide the necessary slope or hydraulic gradient to cause the water to flow at a sufficient rate. So the outlet from the pump to the channel needs to be slightly higher than the field level, thus requiring an increased static head and therefore an increased power demand.

For the same reason the secondary or field distribution system will also create an additional pumping head, either because of pipe friction, or if sprinklers are used then extra pressure is needed to propel the jets of water. Even open channels or furrows imply extra static head because of the need to allow for water to flow downhill.

The power needed is the product of head and flow, and any losses that cause water to fail to reach the plants also represent a reduction in effective flow from the system. Such losses therefore add to the power demand and represent a further source of inefficiency. Typical water losses are due to leakage from the conveyance system before reaching the field, evaporation and percolation into the soil away from crop roots.

Therefore, in common with the prime-mover and the water lifting device an entire irrigation system can be sub-divided into stages, each of which has a (variable) efficiency and a discrete need for power, either through adding to the actual pumping head or through decreasing the effective flow rate due to losses of water (or both).

Most components have an optimum efficiency. In the case of passive items like pipes or distribution systems this might be redefined as "cost-effectiveness" rather than mechanical efficiency. All components need to be chosen so as to be optimized close to the planned operating condition of the system if the most economical and efficient system is to be derived. The concept of "cost-effectiveness" is an important one in this connection, since most irrigation systems are a compromise or trade-off between the conflicting requirements of minimizing the capital cost of the system and minimizing the running costs. This point may be illustrated by a comparison between earth channels and aluminium irrigation pipes as a conveyance; the channels are usually cheap to build but require regular maintenance, offer more resistance to flow and, depending on the soil conditions, are prone to lose water by both percolation and evaporation. The pipe is expensive, but usually needs little or no maintenance and involves little or no loss of water.

Because purchase costs are obvious and running costs (and what causes them) are less clear, there is a tendency for small farmers to err on the side of minimizing capital costs. They also do this as they so often lack finance to invest in a better system. This frequently results in poorer irrigation system efficiencies and reduced returns then may be possible with a more capital-intensive but better optimized system.

2.1.4 Flow Through Channels and Pipes

The proper design of water conveyance systems is complex, and numerous text-books deal with this topic in detail. It is therefore only proposed here to provide an outline of the basic principles so far as they are important to the correct choice and selection of water lifting system. Useful references on this subject are [3] and [8].

i. Channels

When water is at rest, the water level will always be horizontal; however, if water flows down an open channel or canal, the water level will slope downwards in the direction of flow. This slope is called the "hydraulic gradient"; the greater the frictional resistance to flow the steeper it will be. Hydraulic gradient is usually measured as the ratio of the vertical drop per given length of channel; eg. 1m per 100m is expressed as 1/100 or 0.01. The rate of flow (Q) that will flow down a channel depends on the cross sectional area of flow (A) and the mean velocity (v). The relationship between these factors is:

$$Q = vA$$

For example, if the cross sectional area is 0.5m², and the mean velocity is 1m/s, then the rate of flow will be:

$$1 \times 0.5 = 0.5 \text{ m}^3/\text{s}$$

The mean velocity (v) of water in a channel can be determined with reasonable accuracy for typical irrigation channels by the Chezy Formula:

$$V = C\sqrt{rs}$$

where C is the Chezy coefficient which is dependent on the roughness of the surface of the channel (n), its hydraulic radius (r), (which is the area of cross-section of submerged channel divided by its wetted submerged perimeter), and the hydraulic gradient (s) of the channel (measured in unit fall per unit length of channel).

The Chezy coefficient is found from Manning's Formula:

$$C = K \frac{r^{1/6}}{n}$$

in this formula; K = 1 if metric units are used, or K = 1.486 if feet are used; r is the previously defined hydraulic radius and n is the Manning's Coefficient of Roughness appropriate to the material used to construct the channel, examples of which are given in Table 2. This table is also of interest in that it indicates the recommended side slopes and maximum flow velocities for a selection of commonly used types of channels, ranging from earth ditches to concrete, metal or wooden flumes. Combining the above equations gives an expression for the quantity of water that will flow down a channel under gravity as follows:

$$Q = AK \frac{r^{2/3} s^{1/2}}{n}$$

where Q will be in m³/s, if A is in m², r is in metres, and K is 1.

To obtain a greater flow rate, either the channel needs to be large in cross section (and hence expensive in terms of materials, construction costs and land utilization) or it needs to have a greater slope. Therefore irrigation channel design always introduces the classic problem of determining the best trade-off between capital cost or first cost (i.e. construction cost) and running cost in terms of the extra energy requirement if flow is obtained by increasing the hydraulic gradient rather than the cross sectional area. The nature of the terrain also comes into consideration, as channels normally need to follow the natural slope of the ground if extensive regrading or supporting structures are to be avoided.

Obviously in reality, the design of a system is complicated by bends, junctions, changes in section, slope or surface, etc. The reader wishing to study this topic in greater detail should refer to an appropriate text book on this subject.

A further point to be considered with channels is the likely loss of water between the point of entry to the channel and the point of discharge caused by seepage through the channel walls and also by evaporation from the open surface. Any such losses need to be made up by extra inputs of water, which in turn require extra pumping power (and energy) in proportion. Seepage losses are of course most significant where the channel is unlined or has fissures which can lose water, while evaporation only becomes a problem for small and medium scale irrigation schemes with channels having a large surface area to depth ratio and low flow rates, particularly under hot and dry conditions; the greatest losses of this kind occur generally within the field distribution system rather than in conveying water to the field. The main factors effecting the seepage rate from a channel or canal are:

i. soil characteristics

ii. depth of water in the channel in relation to the wetted area and the depth of the groundwater

iii. sediment in water in relation to flow velocity and length of time channel has been in use

This latter point is important [9], as any channel will leak much more when it has been allowed to dry out and then refill. Seepage decreases steadily through the season due to sediment filling the pores and cracks in the soil. Therefore, it is desirable to avoid letting channels dry out completely to reduce water losses when irrigating on a cyclic basis.

Typical conveyance efficiencies for channels range at best from about 90% (or more) with a heavy clay surface or a lined channel in continuous use on small to medium land holdings down to 60-80% in the same situation, but with intermittent use of the channel. In less favourable conditions, such as on a sandy or loamy soil, also with intermittent use, the conveyance efficiency may typically be 50-60% or less; (i.e. almost half the water entering the channel failing to arrive at the other end).

Methods for calculating conveyance losses have been derived and are discussed in detail in specialist references (such as [9]). For example, an approach used by the Irrigation Department in Egypt [9] uses an empirical formula attributed to Molesworth and Yennidumia:

$$S = c L P \sqrt{R}$$

where S will be the conveyance loss in m³/s per length L

if
- c = coefficient depending on nature of soil (eg. $c = 0.0015$ for clay and .003 for sand)
- L = length in km
- P = wetted perimeter of cross section
- R = hydraulic mean depth (ie. flow cross-sectional area divided by width of surface)

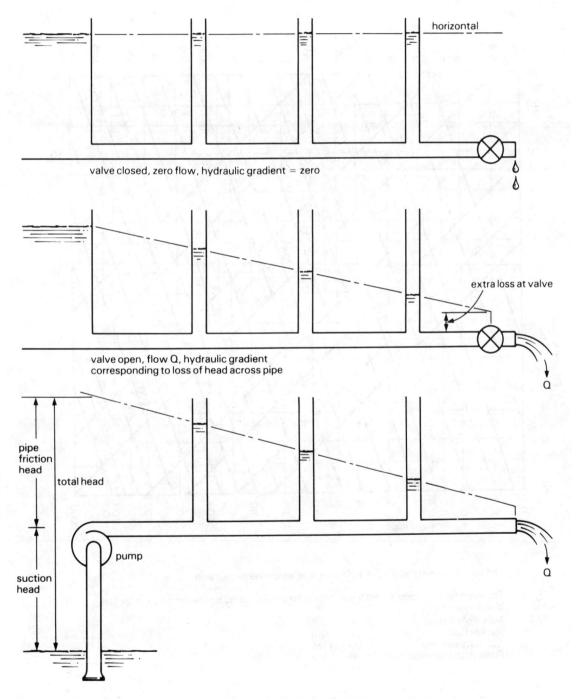

Fig. 3 The concept of an 'hydraulic gradient'

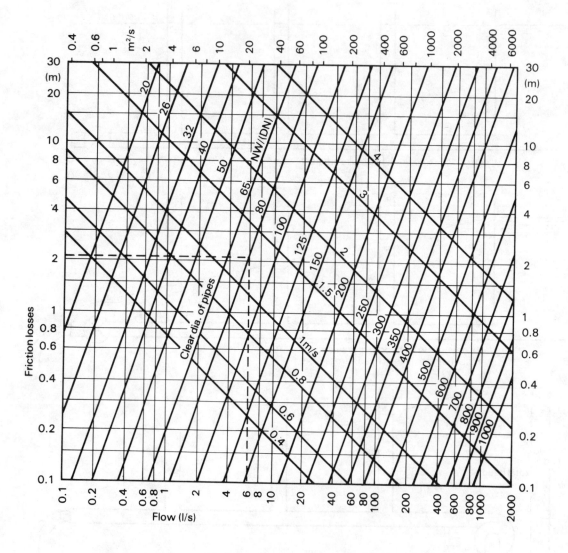

Friction losses in metres per 100m for a new pipeline of cast iron

For other types of pipe multiply the friction loss as indicated by the table by the factors given below:

New Rolled steel	–	0.8
New plastic	–	0.8
Old rusty cast iron	–	1.25
Pipes with encrustations	–	1.7

Fig. 4   Determination of head friction losses in straight pipes

ii.   Pipes

A pipe can operate like a channel with a roof on it; i.e. it can be unpressurized, often with water not filling it. The advantage of a pipe, however, is that it need not follow the hydraulic gradient like a channel, since water cannot overflow from it if it dips below the natural level. In other words, although pipes are more expensive than channels in relation to their carrying capacity, they generally do not require accurate levelling and grading and are therefore more cheaply and simply installed. They are of course essential to convey water to a higher level or across uneven terrain. As with a channel, a pipe also is subject to a hydraulic gradient which also necessarily becomes steeper if the flow is increased; in other words a higher head or higher pressure is needed to overcome the increased resistance to a higher flow. This can be clarified by imagining a pipeline with vertical tappings of it (as in Fig. 3). When no flow takes place due to the outlet valve being closed, the water pressure along the pipe will be uniform and the levels in the vertical tappings will correspond to the head of the supply reservoir. If the valve is opened so that water starts to flow, then a hydraulic gradient will be introduced as indicated in the second diagram and the levels in the vertical tappings will relate to the hydraulic gradient, in becoming progressively lower further along the pipe. The same applies if a pump is used to push water along a pipe as in the lowest diagram in the figure. Here the pump needs to overcome a resistance equal to the static head of the reservoir indicated in the two upper diagrams, which is the pipe friction head. In low lift applications, as indicated, the pipe friction head can in some cases be as large or larger then the static head (which in the example is all suction head since the pump is mounted at the same level as the discharge). The power demand, and hence the energy costs will generally be directly related to total head for a given flow rate, so that in the example, friction losses in the pipe could be responsible for about half the energy costs.

Those wishing to undertake scientifically rigorous analysis should consult a specialized hydraulics text book, (eg. [8] or [9]) but an approximate value of the head loss through a pipe can be gained using the empirical equation [8],[10]:

$$H_f = K \frac{LQ^2}{C^2 D^5}$$

where the loss of head, H_f is calculated using:
L = length of pipe
Q = flow rate
C = coefficient of friction for pipe
D = internal diameter of pipe

The head loss due to friction is expressed as an "hydraulic gradient", i.e. head per length of pipe (m per m or ft per ft).

Note: use K = 10 with metric units, (L and D in metres and Q in cubic metres per second), and K = 4.3 with L and D in feet and Q in cubic feet per second. Values of C are typically 1.0 for steel, 1.5 for concrete, 0.8 for plastics.

An easy way to estimate pipe friction is to use charts, such as Fig. 4. Reference to this figure indicates that a flow, for example, of

6 litres/second (95 US gall/minute) through a pipe of 80mm (3" nominal bore) diameter results in a loss of head per 100m of pipe of just over 2m. As an alternative method, Fig. 5 gives a nomogram (from reference [10]) for obtaining the head loss, given in this case as m/km, for rigid PVC pipe. These results must be modified, depending on the type of pipe, by multiplying the result obtained from the chart by the roughness coefficient of the pipe relative to the material for which the chart of nomogram was derived; for example, if Fig. 4 is to be used for PVC pipe, the result must be multiplied by the factor 0.8 (as indicated at the foot of the figure) because PVC is smoother than iron and typically therefore imposes only 80% as much friction head.

Account must also be taken of the effects of changes of cross section, bends, valves or junctions, which all tend to create turbulence which in effect raises the effective friction head. Ageing of pipes due to growth of either organic matter or corrosion, or both, also increases the friction head per unit length because it increases the frictional resistance and it also decreases the available cross section of flow. This is a complex subject and various formulae are given in text books to allow this effect to be estimated when calculating head losses in pipes.

The head loss due to friction in a pipeline is approximately related to the mean velocity and hence the flow rate squared; i.e.:

$$\text{head loss} \quad h_f = KQ^2$$

therefore, the total head felt by a pump will be approximately the sum of the static head, the friction head and (if the water emerges from the outlet with significant velocity) the velocity head:

$$\text{total head} \quad H_t = h_s + KQ^2 + \frac{V^2}{2g}$$

i.e. (total head) = (static head) + (friction head) + (velocity head)

Since the velocity of flow is proportional to the flow rate (Q), the above equation can be re-written:

$$\text{total head} \quad H_t = h_s + K'Q^2$$

where

$$K' = K + \frac{1}{2gA^2}$$

Fig. 6 illustrates the relationship between the total head and the flow rate for a pumped pipeline, and the pipeline efficiency which can be expressed in energy terms as:

$$\text{pipeline efficiency} \quad \eta_{pipe} = \frac{h_s - K'Q^2}{h_s}$$

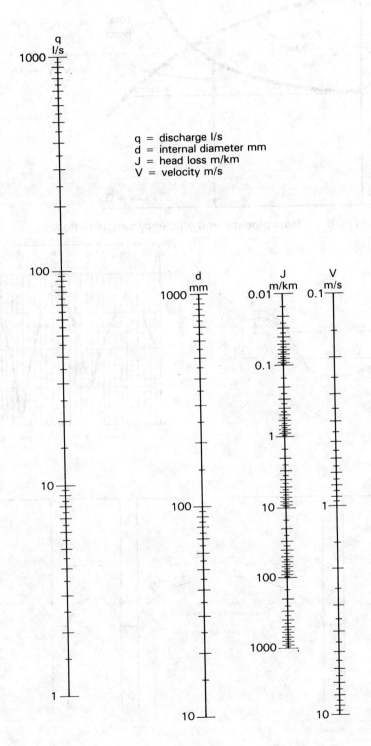

Fig. 5   Head loss nomogram calculated for rigid PVC pipes using Blasius formula

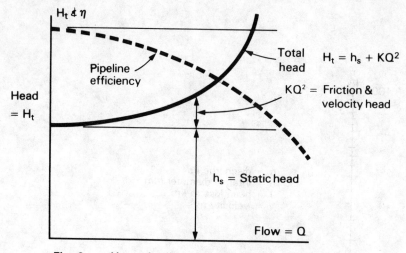

Fig. 6   How pipeline and efficiency vary with flow

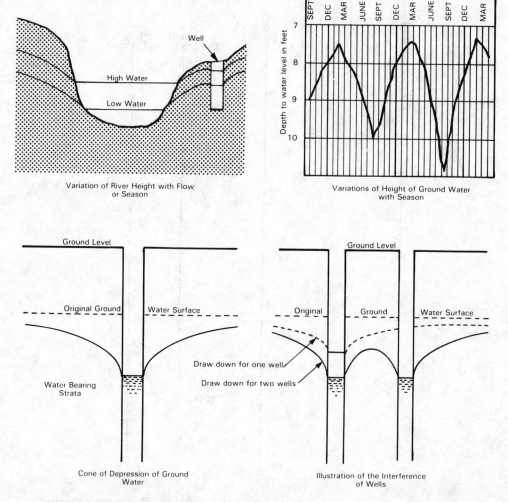

Fig. 7   Effects of various physical conditions on the elevation of water surfaces in wells

2.1.5 Suction Lift: the Atmospheric Limit

Certain types of pump are capable of sucking water from a source; i.e. the pump can be located above the water level and will literally pull water up by creating a vacuum in the suction pipe. Drawing water by suction depends on the difference between the atmospheric pressure on the free surface of the water and the reduced pressure in the suction pipe developed by the pump. The greater the difference in pressure, the higher the water will rise in the pipe. However, the maximum pressure difference that can be created is between sea level atmospheric pressure on the free surface and a pure vacuum, which theoretically will cause a difference of level of water of 10.4m (or 34ft). However, before a drop in pressure even approaching a pure vacuum can be produced, the water will start gassing due to release of air held in solution (just like soda water gasses when released from a pressurized container); if the pressure is reduced further, the water can boil at ambient temperature. As soon as this happens, the pump loses its prime and the discharge will cease (due to loss of prime) or at least be severely reduced. In addition, boiling and gassing within the pump (known as cavitation) can cause damage if allowed to continue for any length of time.

The suction lifts that can be achieved in practice are therefore much less than 10.4m. For example, centrifugal pumps, which are prone to cavitation due to the high speed of the water through the impeller, are generally limited to a suction lift of around 4.5m (15ft) even at sea level with a short suction pipe. Reciprocating pumps generally impose lower velocities on the water and can therefore pull a higher suction lift, but again, for practical applications, this should never normally exceed about 6.5m (21ft) even under cool sea level conditions with a short suction pipe.

At higher altitudes, or if the water is warmer than normal, the suction lift will be reduced further. For example, at an altitude of 3,000m (10,000ft) above sea level, due to reduced atmospheric pressure, the practical suction lift will be reduced by about 3m compared with sea level, (and proportionately for intermediate altitudes, so that 1,500m above sea level will reduce suction lift by about 1.5m). Higher water temperatures also cause a reduction in practical suction head; for example, if the water is at say 30°C, (or 86°F) the reduction in suction head compared with water at a more normal 20°C will be about 7%.

Extending the length of the suction pipe also reduces the suction head that is permissible, because pipe friction adds to the suction required; this effect depends on the pipe diameter, but typically a suction pipe of say 80m length will only function satisfactorily on half the above suction head.

2.1.6 Draw-down and Seasonal Variations of Water Level

Groundwater and river water levels vary, both seasonally and in some cases due to the rate of pumping. Such changes in head can significantly influence the power requirements, and hence the running costs. However, changes in head can also influence the efficiency with which the system works, and thereby can compound any extra running costs caused by a head increase. More serious problems can arise, resulting in total system failure, if for example a surface mounted suction pump is in use, and the supply water level falls sufficiently to make the suction lift exceed the practical suction lift limits discussed in the previous section.

Fig. 7 illustrates various effects on the water level of a well in a confined aquifer. The figure shows that there is a natural ground water level (the water table), which often rises either side of a river or pond since ground water must flow slightly downhill into the open water area. The water table tends to develop a greater slope in impermeable soils (due to higher resistance to flow and greater capillary effects), and is fairly level in porous soil or sand.

If a well is bored to below the water table and water is extracted, the level in the well tends to drop until the inflow of water flowing "downhill" from the surrounding water table balances the rate at which water is being extracted. This forms a "cone of depression" of the water table surrounding the well. The greater the rate of extraction, the greater the drop in level. The actual drop in level in a given well depends on a number of factors, including soil permeability and type, and the wetted surface area of well below the water table (the greater the internal surface of the well the greater the inflow rate that is possible). Extra inflow can be gained either by increasing the well diameter (in the case of a hand-dug well) or by deepening it (the best possibility being with a bore-hole).

Draw-down usually will increase in proportion to extraction rate. A danger therefore if large and powerful pumps are used on small wells or boreholes is to draw the water down to the pump intake level, at which stage the pump goes on "snore" (to use a commonly used descriptive term). In other words, it draws a mixture of air and water which in many cases causes it to lose its prime and cease to deliver. As with cavitation, a "snoring" pump can soon be damaged. But not only the pump is at risk; excessive extraction rates on boreholes can damage the internal surface below the water table and cause voids to be formed which then leads to eventual collapse of the bore. Even when a fully lined and screened borehole is used, excessive extraction rates can pull a lot of silt and other fine material out with the water and block the screen and the natural voids in the surrounding sub-soil, thereby increasing the draw-down further and putting an increasing strain on the lowermost part of the bore. Alternatively, with certain sub-soils, the screen slots can be eroded by particles suspended in the water, when the extraction rate is too high, allowing larger particles to enter the bore and eventually the possible collapse of the screen.

Neighbouring wells or boreholes can influence each other if they are close enough for their respective cones of depression to overlap, as indicated in Fig. 7. Similarly, the level of rivers and lakes will often vary seasonally, particularly in most tropical countries having distinct monsoon type seasons with most rain in just a few months of the year. The water table level will also be influenced by seasonal rainfall, particularly in proximity to rivers or lakes with varying levels, (as indicated in Fig. 7).

Therefore, when using boreholes, the pump intake is best located safely below the lowest likely water level, allowing for seasonal changes and draw-down, but above the screen in order to avoid producing high water velocities at the screen.

When specifying a mechanized pumping system, it is therefore most important to be certain of the minimum and maximum levels if a surface water source is to be used, or when using a well or borehole, the draw-down to be expected at the proposed extraction rate. A pumping test is necessary to determine the draw-down in wells and boreholes; this is normally done by extracting water with a portable engine-pump, and measuring the drop in level at various pumping rates after the level has stablized. In many countries, boreholes are normally pumped as a matter of routine to test their draw-down and the information from the pumping test is commonly logged and filed in the official records and can be referred to later by potential users.

2.1.7 Review of a Complete Lift Irrigation System

The factors that impose a power load on a pump or water lifting device are clearly more complicated than simply multiplying the static head between the water source and the field by the flow rate. The load consists mainly of various resistances to flow which when added together comprise the gross pumping head, but it also is increased by the need to pump extra water to make up for losses between the water source and the crop.

Fig. 8 summarizes these in a general way, so that the advantages and disadvantages of different systems discussed later in this paper can be seen in the context of their general efficiency. The table indicates the various heads and losses which are superimposed until the water reaches the field; actual field losses are discussed in more detail in Section 2.2 which follows.

The system hydraulic efficiency can be defined as the ratio of hydraulic energy to raise the water delivered to the field through the static head, to the hydraulic energy actually needed for the amount of water drawn by the pump:

$$\text{system hydraulic efficiency} \quad \eta_{sys} = \frac{E_{stat}}{E_{gross}}$$

Where E_{stat} is the hydraulic energy output, and E_{gross} is hydraulic energy actually applied.

Finally, Fig. 9 indicates the energy flow through typical complete irrigation water lifting and distribution systems and shows the various losses.

2.1.8 Practical Power Requirements

Calculating the power requirement for water lifting is fundamental to determining the type and size of equipment that should be used, so it is worth detailing the principles for calculating it. In general the maximum power required will simply be:

$$\frac{\text{(Maximum mass flow delivered)} \times \text{(static head)} \times g}{\text{(total system efficiency at max. flow)}}$$

where the mass flow is measured in kg/s of water. 1kg of water is equal to 1 litre in volume, so it is numerically equal to the flow in litres per second; g is the acceleration due to gravity of $9.81 m/s^2$ (or $32.2 ft/s^2$). Therefore, for example, 5 litre/sec through 10m with a system having an overall efficiency of 10% requires:

$$\frac{5\,(kg/s) \times 10(m) \times 9.81(m/s^2)}{0.10\,(efficiency)} = 4905W$$

The daily energy requirement will similarly be:

$$\frac{(\text{mass of water delivered per day}) \times (\text{head}) \times g}{(\text{average system efficiency})}$$

eg. for 60m³/day lifted through 6m with an average efficiency of 5%:

$$\frac{60,000 \times 6 \times 9.81}{0.05} = 70.6\,MJ/day \text{ (million joules/day)}$$

Note: 60m3 = 60,000 litres which in turn has a mass of 60,000kg (= 60 tonne). Also, since 1kWh = 3.6MJ, we can express the above result in kWh simply by dividing by 3.6:

so 70.6 MJ/day = 19.6 kWh/day

It follows from these relationships that a simple formula can be derived for converting an hydraulic energy requirement into kWh, as follows:

$$E_{hyd} = \frac{QH}{367}$$

If the above calculation relates to a gasoline engine pump irrigation system, as it might with the figures chosen for the example, then we know that as the energy input is 19.6kWh/day and as gasoline typically has an energy content of 32MJ/litre or 8.9kWh/litre, this system will typically require an input of 2.2 litre of gasoline per day.

Fig. 10 illustrates the hydraulic power requirement to lift water at a range of pumping rates appropriate to the small to medium sized land-holdings this publication relates to. These figures are the hydraulic output power and need to be divided by the pumping system efficiency to arrive at the input power requirement. For example, if a pump of 50% efficiency is used, then a shaft power of twice the hydraulic power requirement is needed; (pump efficiencies are discussed in more detail in Chapter 3). The small table on Fig. 10 indicates the typical hydraulic power output of various prime movers when working with a 50% efficient water lifting device; i.e. it shows about half the "shaft power" capability. The ranges as indicated are meant to show "typical" applications; obviously there are exceptions.

These power curves, which are hyperbolas, make it difficult to show the entire power range of possible interest in connection with land-holdings from less than 1ha to 25ha, even though they cover the flow, head and power range of most general interest. Fig. 11 is a log-log graph of head versus flow, which straightens out the power curves and allows easier estimation of the hydraulic power requirement for flows up to 100 l/s and hydraulic powers of up to 16kW.

Fig. 12 is perhaps more generally useful, being a similar log-log graph, but of daily hydraulic energy requirement to deliver different volumes of water through a range of heads of up to 32m. The area of land that can be covered, as an example to 8mm depth, using a given hydraulic energy output

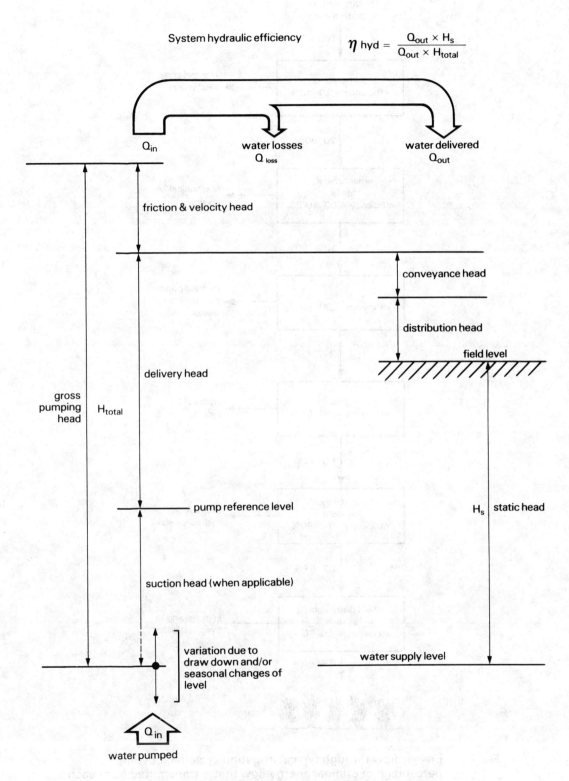

Fig. 8   Factors affecting system hydraulic efficiency

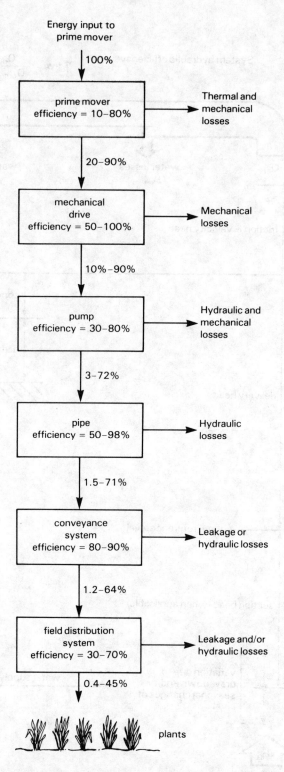

Fig. 9 Energy flow through typical irrigation system (showing percentage of original energy flow that is transmitted from each component to the next).

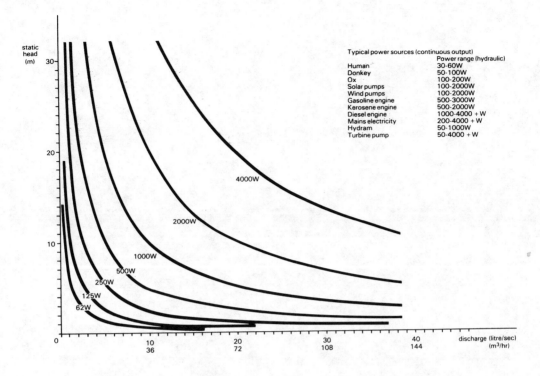

Fig. 10   Hydraulic power requirements to lift water

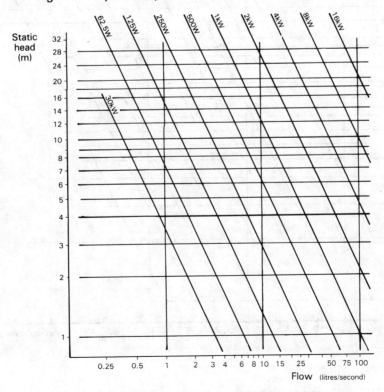

Fig. 11   Relationship between power, head and flow

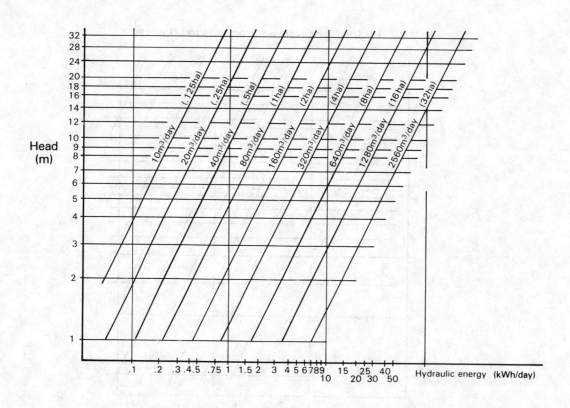

Fig. 12 Relationship between energy, head and daily output (areas that can be irrigated to a depth of 8mm are shown in parentheses)

over the range of heads is also given.

Finally, Fig. 13 is a nomograph which allows the entire procedure of calculating power needs for a given irrigation requirement to be reduced to ruling a few lines so as to arrive at an answer. The following example of the procedure is indicated and helps to illustrate the process; starting with the area to be irrigated (in the example 3ha is used), rule a line vertically upwards until it intersects the diagonal. This point of intersection gives the required depth of irrigation; 8mm is used in the example but field and distribution losses are not accounted for in this nomograph, so the irrigation demand used must be the gross and not the nett requirement. Rule horizontally from the point of intersection, across the vertical axis (which indicates the daily water requirement in cubic metres per day - 240 in the example) until the line intersects the diagonal relating to the pumping head; 10 m head is used in the example. Dropping a vertical line from the point of intersection gives the hydraulic energy requirement (6.5kWh (hyd)/day). This is converted to a shaft energy requirement by continuing the line downwards to the diagonal which corresponds with the expected pumping efficiency; 50% efficiency is assumed for the example (the actual figure depends on the type of pumping system) and this gives a shaft power requirement of 13kWh/day when a line is ruled horizontally through the shaft power axis. The final decision is the time per day which is to be spent pumping the required quantity of water; 5h is used as the example. Hence, ruling a line vertically from the point of intersection to the average power axis (which coincides with the starting axis), shows that a mean power requirement (shaft power) of about 2.6kW is necessary for the duty chosen in the example. It should be noted that this is mean shaft power; a significantly higher peak power or rated power may be necessary to achieve this mean power for the number of hours necessary.

This nomograph readily allows the reader to explore the implications of varying these parameters; in the example it is perhaps interesting to explore the implications of completing the pumping in say 3h rather than 5h and it is clear that the mean power requirement then goes up to about 4.25kW.

In some cases it may be useful to work backwards around the nomograph to see what a power unit of a certain size is capable of doing in terms of areas and depths of irrigation.

The nomograph has been drawn to cover the range from 0 - 10ha, which makes it difficult to see clearly what the answers are for very small land-holdings of under 1ha. However, the nomograph also works if you divide the area scale by 10, in which case it is also necessary to divide the answer in terms of power needed by 10. In the example, if we were interested in 0.3ha instead of 3ha, and if the same assumptions are used on depth of irrigation, pumping head, pump efficiency and hours per day for pumping, the result will be 0.26kW (or 260W) instead of 2.6kW as indicated. Obviously the daily water requirement from the top axis will also need to be divided by ten, and in the example will be 24m^3/day. Similarly, it is possible to scale the nomograph up by a factor of ten to look at the requirements for 10 to 100 ha in exactly the same way. Note that in most real cases, if the scale is changed, factors like the pump efficiency ought to be changed too. An efficiency of 50% used in the example is a poorish efficiency for a pump large enough to deliver 240m^3/day, but it is rather a high efficiency for a pump capable of only one tenth of this daily discharge.

2.2 OUTLINE OF PRINCIPLES OF SMALL SCALE IRRIGATION

2.2.1 Irrigation Water Requirements

The quantity of water needed to irrigate a given land area depends on numerous factors, the most important being:

- nature of crop
- crop growth cycle
- climatic conditions
- type and condition of soil
- topography
- conveyance efficiency
- field application efficiency
- water quality
- effectiveness of water management

Few of these factors remain constant, so that the quantity of water required will vary from day to day, and particularly from one season to the next. The selection of a small scale irrigation system needs to take all of the above factors into account.

The crop takes its water from moisture held in the soil in the root zone. The soil therefore effectively acts as a water storage for the plants, and the soil moisture needs replenishing before the moisture level falls to what is known as the "Permanent Wilting Point" where irreversible damage to the crop can occur. The maximum capacity of the soil for water is when the soil is "saturated", although certain crops do not tolerate water-logged soil and in any case this can be a wasteful use of water. In all cases there is an optimum soil moisture level at which plant growth is maximized (see Fig. 12). The art of efficient irrigation is to try to keep the moisture level in the soil as close to the optimum as possible.

References such as [3], [8], [10] and [11] give a more detailed treatment of this subject.

2.2.2 Nett Irrigation Requirement

The estimation of irrigation water requirements starts with the water needs of the crop. First the "Reference Crop Evapotranspiration" ET_o is determined; this is a standardized rate of evapotranspiration (related to a reference crop of tall green grass completely shading the ground and not short of water) which provides a base-line and which depends on climatic factors including pan evaporation data and windspeed. A full description on the determination of ET_o is presented in reference [11]. Because ET_o depends on climatic factors, it varies from month to month, often by a factor of 2 or more. The evapotranspiration of a particular crop (ET_{crop}) will of course be different from that of the reference crop, and this is determined from the relationship:

$$ET_{crop} = ET_o \times K_c$$

K_c is a "crop coefficient" which depends on the type of crop, its stage of growth, the growing season and the prevailing climatic conditions. It can vary typically from around 0.3 during initial growth to around 1.0 (or a bit over 1.0) during the mid-season maximum rate of growth period; Fig. 13 shows an example. Therefore the actual value of the crop water requirement, ET_{crop}, usually varies considerably through the growing season.

The actual nett irrigation requirement at any time is the crop evapotranspiration demand, minus any contributions from rainfall, groundwater or stored moisture in the soil. Since not all rainfall will reach the plant roots, because a proportion will be lost through run-off, deep percolation and evaporation, the rainfall is factored to arrive at a figure for "effective rainfall". Also, some crops require water for soil preparation, particularly for example, rice, and this need has to be allowed for in addition to the nett irrigation requirement.

To give an idea of what these translate into in terms of actual water requirements an approximate "typical" nett irrigation requirement under tropical conditions with a reasonably efficient irrigation system and good water management is 4,000m³/ha per crop, but under less favourable conditions as much as 13,000m³/ha per crop can be needed. This is equal to 400-1300mm of water per crop respectively. Since typical growing cycles are in the range of 100-150 days in the tropics, the average daily requirement will therefore be in the 30-130m³/ha range (3-13mm/day). Because the water demand varies through the growing season, the peak requirement can be more than double the average, implying that a nett peak output of 50-200m³/ha will generally be required (which gives an indication of the capacity of pumping system needed for a given area of field).

Inadequate applications of irrigation water will not generally kill a crop, but are more likely to result in reduced yield [11]. Conversely, excessive applications of water can also be counterproductive apart from being a waste of water and pumping energy. Accurate application is therefore of importance mainly to maximise crop yields and to get the best efficiency from an irrigation system.

2.2.3 Gross Irrigation Requirement

The output from the water lifting device has to be increased to allow for conveyance and field losses; this amount is the gross irrigation requirement. Typical conveyance and field distribution system efficiencies are given in Table 3 [11] [12], from which it can be seen that conveyance efficiencies fall into the range 65-90% (depending on the type of system), while "farm ditch efficiency" or field application efficiency will typically be 55-90%. Therefore, the overall irrigation system efficiency, after the discharge from the water lifting device, will be the product of these two; typically 30-80%. This implies a gross irrigation water requirement at best about 25% greater than the nett requirement for the crop, and at worst 300% or more.

The previous "typical peak nett irrigation" figures of 50-200m³/day per hectare imply "peak gross irrigation" requirements of 60-600m³/day; a wide variation due to compounding so many variable parameters. Clearly there is often much scope for conservation of pumping energy by improving the water distribution efficiency; investment in a better conveyance and field distribution system will frequently pay back faster than investment in improved pumping capacity and will achieve the same result. Certainly costly pumping systems should generally only be considered in conjunction with efficient conveyance and field distribution techniques. The only real justification for extravagant water losses is where pumping costs are low and water distribution equipment is expensive.

A summary of the procedure so far outlined to arrive at the gross crop irrigation water requirement is given in Fig. 14.

2.2.4 Pumping Requirement

In order to specify a water lifting system the following basic information is needed:

i. the average water demand through the growing season

ii. the peak daily water demand (which generally will occur when the crop coefficient and rate of plant growth are at their peak)

Having determined the daily application required by the plants, a further consideration is the "intake rate" as different soil types absorb water at different rates, (see Table 4). Too rapid a rate of application on some soils can cause flooding and possible loss of water through run-off. This constraint determines the maximum flow rate that can usefully be absorbed by the field distribution system. For example, some silty clay soils can only take about 7 l/sec per hectare, but in contrast sandy soils do not impose a serious constraint as they can often usefully absorb over 100 l/s per hectare. Obviously lower rates than the maximum are acceptable, although the application efficiency is likely to be best at a reasonably high rate in most cases, and farmers obviously will prefer not to take longer than necessary to complete the job.

Taking account of the above constraint on flow rate, it is then possible to calculate how many hours per day the field will require irrigating, for example by using the nomogram given in Fig. 15.

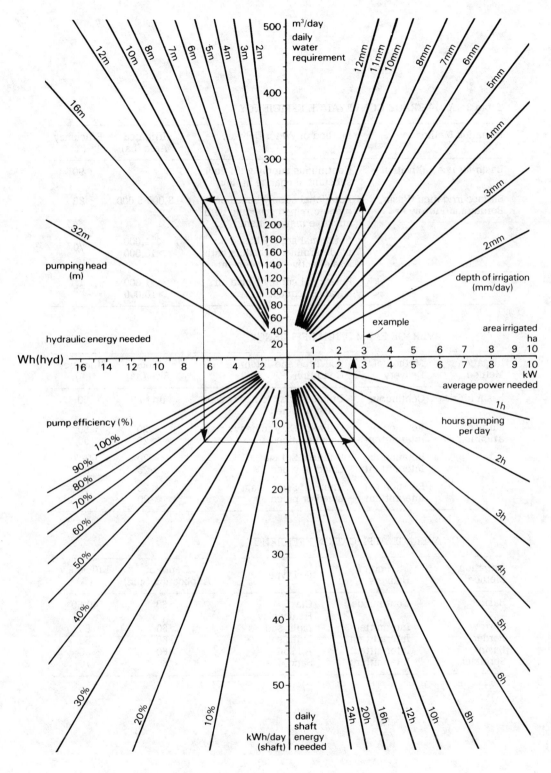

Fig. 13 Nomogram for calculating power needs for a given area, depth of irrigation and head

Table 3 A AVERAGE CONVEYANCE EFFICIENCY

Irrigation Method	Method of Water Delivery	Irrigated Area (ha)	Efficiency (%)
Basin for rice cultivation	Continuous supply with no substantial change in flow	—	90
Surface irrigation (Basin, Border and Furrow	Rotational supply based on predetermined schedule with effective management	3,000-5,000	88
	Rotational supply based on predetermined schedule with less effective management	< 1,000 >10,000	70
	Rotational supply based on advance request	< 1,000 >10,000	65

B AVERAGE FARM DITCH EFFICIENCY

Irrigation Method	Method of Delivery	Soil Type and Ditch Condition	Block Size (ha)	Efficiency (%)
Basin for rice	Continuous	Unlined: Clay to heavy clay Lined or piped	up to 3	90
Surface Irrigation	Rotation or Intermittent	Unlined: Clay to heavy clay Lined or piped	<20 >20	80 90
	Rotation or Intermittent	Unlined: Silt clay Lined or piped	<20 >20	60-70 80
	Rotation or Intermittent	Unlined: Sand, loam Lined or piped	<20 >20	55 65

C AVERAGE APPLICATION EFFICIENCY

Irrigation Method	Method of Delivery	Soil type	Depth of Application (mm)	Efficiency (%)
Basin	Continuous	Clay Heavy clay	>60	40-50
Furrow	Intermittent	Light soil	>60	60
Border	Intermittent	Light soil	>60	60
Basin	Intermittent	All soil	>60	60
Sprinkler	Intermittent	Sand, loam	<60	70

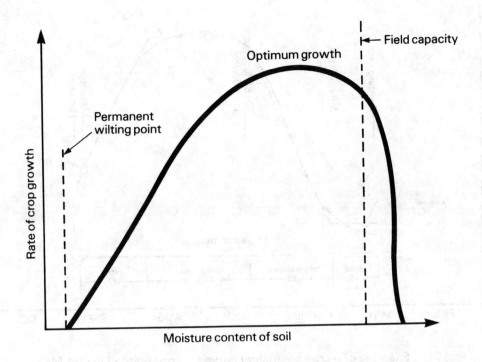

Fig. 14 Rate of crop growth as a function of soil moisture content

Table 4 AVERAGE INTAKE RATES OF WATER IN mm/hr FOR DIFFERENT SOILS AND CORRESPONDING STREAM SIZE l/sec/ha

Soil Texture	Intake Rate mm/hr	Stream size q l/sec/ha
Sand	50 (25 to 250)	140
Sandy loam	25 (15 to 75)	70
Loam	12.5 (8 to 20)	35
Clay loam	8 (2.5 to 5)	7
Silty clay	2.5 (0.03 to 5)	7
Clay	5 (1 to 15)	14

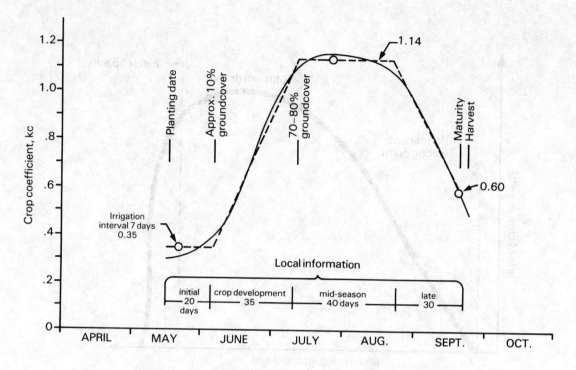

Fig. 15   Example of a crop coefficient curve for corn planted in mid-May at Cairo, Egypt; e.g. initial stage is 8.4mm/day with irrigation frequency of 7 days

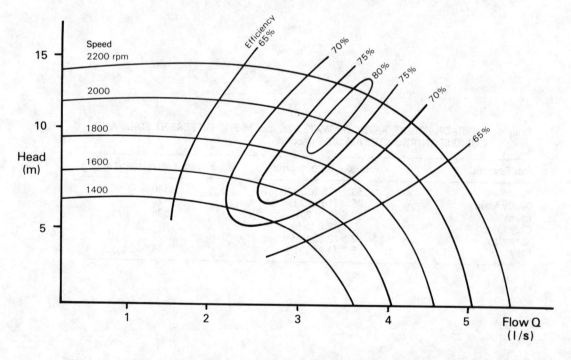

Fig. 16   Typical curves showing relationship between head, flow, speed and efficiency (example given for a centrifugal pump)

Table 5 TAXONOMY OF PUMPS AND WATER LIFTS

Category and Name	Construction	Head range (M)	Power range (W)	Output	Efficiency	Cost	Suction Lift?	Status for Irrigation
I DIRECT LIFT DEVICES								
Reciprocating/cyclic								
Watering can	1	>3	*	*	*	*	×	√
Scoops and bailers	1	>1	*	**	*	*	×	√
Swing basket	1	>1	*	**	*	*	×	√
Pivoting gutters and "dhones"	2	1-1.5	*	**	**	**	×	√
Counterpoise lift or "Shadoof"	2	1-4	*	**	**	**	×	√
Rope & bucket and windlass	1	5-50	**	*	*	*	×	√
Self-emptying bucket or "mohte"	2	3-8	**	***	*	**	×	√
Reciprocating bucket hoist	3	100-500	****	****	***	*****	×	×
Rotary/continuous								
Continuous bucket pump	2	5-50	**	**	***	**	×	√
Persian wheel or "tablia"	2	3-10	**	***	***	**	×	√
Improved Persian wheel "zawaffa"	2	3.15	***	****	****	***	×	√
Scoop wheels or "sakia"	2	>2	**	****	****	****	×	√
Waterwheels or "noria"	2	>5	*	**	**	**	×	√
II DISPLACEMENT PUMP								
Reciprocating/cyclic								
Piston/bucket pumps	2 & 3	2-200	***	***	*****	****	√	√
Plunger pumps	3	100-500	***	**	****	*****	√	?
Diaphragm pumps	3	5-10	**	***	****	***	√	?
"Petropump"	3	10-100	**	**	*****	****	√	?
Semi-rotary pumps	3	5-10	*	**	**	**	×	?
Gas or vapour displacement	3	5-50	****	****	***	***	√ or ×	?
Rotary/continuous								
Gear and lobe pumps	3	10-20	*	*	**	***	√	×
Flexible vane pumps	3	10-20	**	***	***	****	√	×
Progressive cavity (Mono)	3	10-100	***	***	****	****	×	?
Archimedean screw	3	>2m	**	****	***	***	×	√
Open screw pumps	3	>6m	****	*****	****	*****	×	√
Coil and spiral pumps	2	>6m	**	**	***	***	×	√
Flash-wheels & treadmills	2 & 3	>2m	**	****	**	**	×	√
Water-ladders "Dragon spines"	2	>2m	**	***	***	***	×	√
Chain (or rope) and washer	2 & 3	3-20m	***	***	****	****	×	√
Peristaltic pump	3	>3m	*	*	***	***	√	×
Porous rope	3	3-10m	**	**	?	?	×	?
III VELOCITY PUMPS								
Reciprocating/cyclic								
Inertia and "joggle" pumps	2 & 3	2-4	*	**	****	**	×	√
Flap valve pump	1 & 2	2-4	*	*	**	*	×	?
Resonating joggle pump	2	2-10	**	****	****	***	×	?
Rebound inertia	3	2-60	**	*	****	***	√	×
Rotary/continuous								
Propeller (axial flow) pumps	3	5-3	****	*****	****	****	×	√
Mixed flow pumps	3	2-10	****	*****	****	****	×	√
Centrifugal (volute) pumps	3	3-20+	*****	*****	****	***	√	√
Centrifugal (turbine) pumps	3	3-20+	*****	*****	****	****	√	√
Centrifugal (regenerative) pumps	3	10-30	***	***	***	*****	√	×
Jet pumps (water, air or stream)	3	2-20	***	***	***	***	×	×
IV BUOYANCY PUMPS								
Air lift	3	5-50	**	***	**	****	×	×
V IMPULSE PUMPS								
Hydraulic ram	3	10-100	**	**	***	***	×	√
VI GRAVITY DEVICES								
Syphons	1, 2 & 3	1-(−10)	—	*****	—	**	—	√
Qanats or foggara	2	—	—	**	*****	—	—	

Construction: 1 Basic   2 Traditional   3 Industrial

Very low *   Medium-high ****   Yes √   Yes √
Low-medium **   High *****   No ×   Possible ?
Medium ***   Unlikely ×

Fig. 17   The scoop used as a simple hand tool

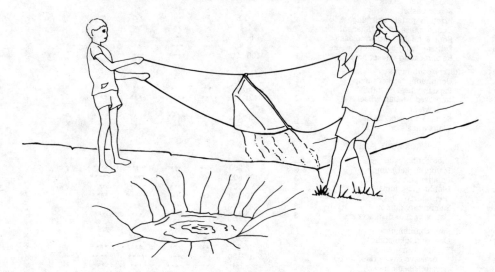

Fig. 18   The swingbasket in use (after T. Schioler [24])

Fig. 19   Scoop with a rope support

3. REVIEW OF PUMPS AND WATER LIFTNG TECHNIQUES

3.1 PRINCIPLES FOR LIFTING AND MOVING WATER

Water may be moved by the application of any one (or any combination) of six different mechanical principles, which are largely independent, i.e. by:

i. <u>direct lift</u>
this involves physically lifting water in a container

ii. <u>displacement</u>
this involves utilizing the fact that water is (effectively) incompressible and can therefore be "pushed" or displaced

iii. <u>creating a velocity head</u>
when water is propelled to a high speed, the momentum can be used either to create a flow or to create a pressure

iv. <u>using the buoyancy of a gas</u>
air (or other gas) bubbled through water will lift a proportion of the water

v. <u>gravity</u>
water flows downward under the influence of gravity

3.2 TAXONOMY OF WATER LIFTS AND PUMPS

Families of pumps and lifting/propelling devices may be classified according to which of the above principles they depend on. Table 5 is an attempt to classify pumps under the categories given above. It will be seen that most categories sub-divide into the further classifications "reciprocating/cyclic" and "rotary". The first of these relates to devices that are cycled through a water-lifting operation (for example a bucket on a rope is lowered into the water, dipped to make it fill, lifted, emptied and then the cycle is repeated); in such cases the water output is usually intermittent, or at best pulsating rather than continuous. Rotary devices were generally developed to allow a greater throughput of water, and they also are easier to couple to engines or other types of mechanical drive. Therefore, by definition, a rotary pump will generally operate without any reversal or cessation of flow, although in some cases the output may appear in spurts or pulsations.

Before considering the differences between the diverse options available for lifting water, it is worth briefly noting the factors they all have in common. Virtually all water lifting devices can best be characterized for practical purposes by measuring their output at different heads and speeds. Normally the performance of a pump is presented on a graph of head versus flow (an H-Q graph, as in Fig. 16) and in most cases curves can be defined for the relationship between H and Q at different speeds of operation. Invariably there is a certain head, flow and speed of operation that represents the optimum efficiency of the device, i.e. where the output is maximized in relation to the power input. Some devices and pumps are more sensitive to variations in these factors than others; i.e. some only function well close to a certain design condition of speed, flow and head, while others can tolerate a wide range of operating conditions with little loss of efficiency. For example, the centrifugal pump characteristic given in Fig. 16 shows an optimum efficiency exceeding 80% is only possible for speeds of about 2,000rpm.

The rest of this Section describes in some detail each of the devices given in the Taxonomy of Pumps and Water Lifting Devices of Table 5.

3.3 RECIPROCATING AND CYCLIC DIRECT LIFT DEVICES

3.3.1 Watering cans, Buckets, Scoops, Bailers and the Swing-Basket

These are all variations on the theme of the bucket, are hand-held and must be the earliest artificial methods for lifting and carrying water. The watering can is effectively a bucket with a built-in sprinkler and represents an efficient, but labour intensive method for irrigating very small land-holdings. Artisan-made watering cans are quite widely used in Thailand. Scoops, bailers (Fig. 17) and the swing-basket (Fig. 18) represent methods of speeding up the process of filling, lifting and emptying a bucket; the latter also uses two people rather than one and thereby increases the mass of water that can be scooped in each swing. These are more fully described in Section 4.2 in the context of using human muscle power, since they have evolved in such a way as to fit the human prime-mover but to be unsuitable for any kind of mechanization. They are rather inefficient as water is lifted over 1m and allowed to fall back to 0.3-0.5m, which is the approximate operating head for devices of this kind.

3.3.2 Suspended Scoop, Gutters, Dhones (or Doons) and the Counterpoise-lift or Shadoof

The next stage of technical advance is to support the mass of water being lifted by mounting the scoop or bucket on a suspended pivoted lever to produce a swinging scoop (Fig. 19) or a see-sawing gutter or "dhone" (Fig. 20) which also operate through low lifts (0.5-1m) at relatively high speed. The water container can be balanced with a weight; Fig. 21 shows a counterpoise lift, alias water-crane or "shadoof".

If the terrain permits, such as on a sloping river bank, several "shadoofs" can be used to lift water in stages through a greater height than is possible with one.

3.3.3 Bucket hoists, Windlasses, Mohtes and Water Skips

To increase the lift it becomes necessary to introduce a rope to pull the container of water from the source to a level where it can be tipped into the conveyance channel. There is therefore a family of devices for pulling up a container of water on a rope. The simplest form for this is a rope and bucket, which in an improved form has a simple windlass, i.e. a hand operated winch, to increase the leverage and hence the size of bucket that can be lifted.

The output of such systems is generally too small for irrigation, (they tend to be used mainly for domestic or livestock water supply duties), but by

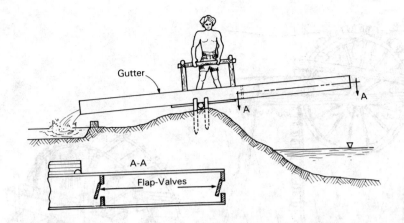

Fig. 20   Dhone as used in Bangladesh

Fig. 21   Counterpoise lift

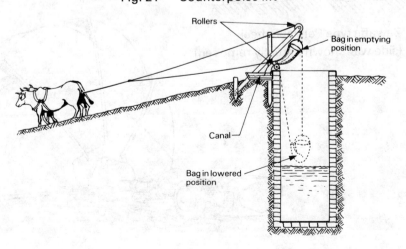

Fig. 22   Self-emptying mohte with inclined tow path

Fig. 23 Persian wheel

Fig. 24 Noria

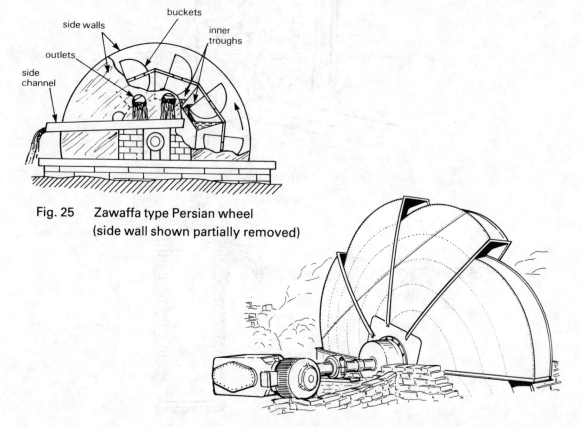

Fig. 25 Zawaffa type Persian wheel
(side wall shown partially removed)

Fig. 26 Sakia or Tympanum (electrically powered in this case)

powering the device with animals, usually oxen, sufficient water can be lifted to irrigate even through heads of 5-10m. This encouraged the evolution of the "self-emptying bucket", known in India as a "mohte", (Figs. 22 and 92). These commonly have a bucket made of leather or rubber, with a hole in its bottom which is held closed by a flap which is pulled tight by a second rope harnessed to the animals. The number in use today is still in the order of a million or more, so this device remains of considerable importance in some regions. Mohtes are discussed in more detail under Section 4.3 dealing with animal power as a prime mover.

3.4 ROTARY DIRECT LIFT DEVICES

It generally improves both efficiency and hence productivity if the water lifting element can move on a steady circular path rather than being cycled or reciprocated. The reason for this is that the energy input to any water lifting device is usually continuous, so that if the output is intermittent, unless energy can be stored during the parts of the cycle when no water is lifted, it is lost. Therefore reciprocating/cyclic devices tend to be less efficient than rotary devices; this is not a firm rule however, as some reciprocating devices include means to store energy through the non-productive part of the cycle while some rotary devices are less efficient for other reasons.

3.4.1 Bucket elevators, Persian wheels and Norias

An obvious improvement to the simple rope and bucket is to fit numerous small buckets around the periphery of an endless belt to form a continuous bucket elevator. The original version of this, which is ancient in origin but still widely used, was known as a "Persian wheel" (Figs. 23, 94 and 95); the earliest forms consisted of earthenware pots roped in a chain which is hung over a drive wheel. The water powered "noria" (Figs. 24, 151 and 152), a water wheel with pots, buckets or hollow bamboo containers set around its rim, is similar in principle except the containers are physically attached to the drive wheel circumference rather than to an endless belt suspended from it.

The flow with any of these devices is a function of the volume of each bucket and the speed at which the buckets pass across the top of the wheel and tip their contents into a trough set inside the wheel to catch the output from the buckets. Therefore, for a given power source and speed of operation roughly the same number of containers are needed regardless of head. In other words, a higher head Persian wheel requires the buckets to be proportionately more spaced out; double the head and you more or less need to double the spacing.

The Persian wheel has been, and still is, widely used particularly in the north of the Indian sub-continent and is discussed in more detail under Section 4.3 on Animal Power, while the noria was widely used in China, S.E. Asia and to some extent in the Middle East and being normally water powered is discussed in more detail in Section 4.9. Both devices are tending to be replaced by more modern mechanical water lifting techniques as they are old-fashioned and low in output. It should be noted that the term "Persian wheel" is sometimes used to describe other types of animal powered rotary pumps.

Although Persian wheels and norias are mechanically quite efficient, the main source of loss from these types of device is that some water is spilled from the buckets and also there is a certain amount of friction drag caused when the buckets scoop up water, which again reduces efficiency. Also, the Persian wheel is obliged to lift the water at least 1 m (or more) higher than necessary before discharging it into a trough, which can significantly increase the pumping head, particularly in the case of low lifts. The traditional wooden Persian wheels also inevitably need to be quite large in diameter to accommodate a large enough collection trough to catch most of the water spilling from the pots; this in turn requires a large well diameter which increases the cost.

Some performance figures for animal powered Persian wheels are given in Molenaar; [1].

Height lifted	Discharge
9m	8-10 m^3/h
6m	10-12 m^3/h
3m	15-17 m^3/h
1.5m	20-22 m^3/h

Depending on the assumption used for the power of the animals in the above examples, the implication is that the efficiency is in the order of 50% at medium lifts such as 6m and perhaps marginally better at higher lifts but worse at lower lifts.

The water-powered noria uses the same principle as the Persian wheel and therefore also needs to be of larger diameter than the pumping head, which either limits it to very low lift pumping or requires very large, cumbersome and expensive construction. Small, low-lift norias, used in Thailand and China, are very inexpensive while much larger norias are used in Vietnam and Syria. Some of the largest in Syria exceed 10m in diameter, but in relation to their size they tend to be unproductive compared with more modern pumping systems. A fuller description of Vietnamese and other norias is given in Section 4.9

3.4.2 Improved Persian wheel, (Zawaffa or Jhallar)

Traditional wooden Persian wheels were fitted with earthenware water containers, but a variety of all-metal improved Persian wheels have been built, some as commercial products, in China, India, Pakistan and Egypt. Metal Persian wheels can be made smaller in diameter, which reduces the extra height the water needs to be lifted before it is tipped out of the containers, and also reduces the well diameter that is necessary.

A modified version of the Persian wheel used in Syria and also in Egypt (where it is called the zawaffa or jhallar) includes internal buckets within the drive wheel which catch the water and direct it through holes in the side plate near the hub into a collection trough; Fig. 25. This reduces both splashing and spillage losses and the extra height above the collection channel at which the water is tipped. Roberts & Singh [13] gave figures for a modernised metal Persian wheel, of 153m^3/h lifted through 0.75m. This implies that efficiencies as high as 75% are possible with modernised devices, which are rather good.

3.4.3 Scoop-wheels; Sakia, Tympanum or Tablia

The scoop-wheel (sakia in Egypt where it originated) has some factors in common with the noria. Although widely used in Egypt it has failed to become

Fig. 27 The Fathi is the optimum design of Sakia

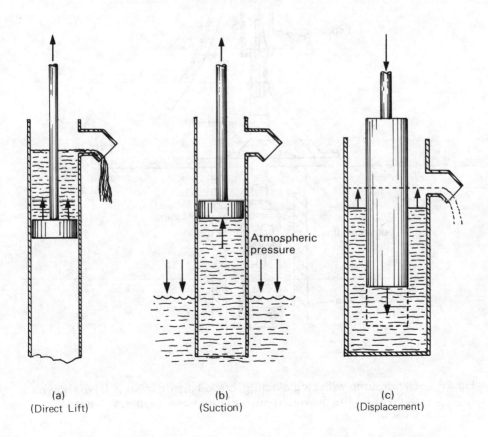

(a) (Direct Lift) (b) (Suction) (c) (Displacement)

Fig. 28 Basic principles of positive displacement pumps

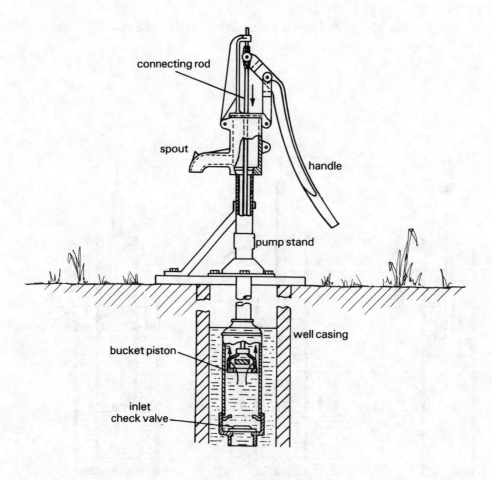

Fig. 29 Hand-pump with single-acting, bucket piston (piston valve shown open as on the down-stroke, and foot valve or inlet valve is closed)

popular anywhere else. It is however an efficient and effective device; Fig. 26.

It consists of a large hollow wheel with scoops around its periphery, which discharges water at or near its hub rather than from its top. The diameters for sakias range from about 2-5m; since water discharges at their hub level, the rule of thumb used in Egypt is that a sakia will lift water through a head of half its diameter less 0.7 m, to allow for the depth of submergence of the rim in order to scoop up water effectively. Therefore sakias of diameters from 2-5m will lift water from 0.3-1.8m respectively.

Sakias are now normally made from galvanized sheet steel. Second-hand vehicle roller bearings are commonly used to support the substantial weight of a sakia and its water contents. Most sakias are animal powered, but they are increasingly being driven by either mains electric motors or small engines, via suitable reduction gearing. The normal operating speed is 2-4 rpm for animal-driven sakias, and 8-15 rpm for motorised or engine-powered units.

Various different spiral shapes have evolved for the internal baffle plates in the sakia, and the Hydraulic Research and Experimental Station (HRES) in Egypt tested various models to try to determine the optimum design. The best shaped designs, such as in Fig. 27, were measured as being as much as 50% better than the worst. Since some 300,000 sakias are in use in the Nile valley and Delta, optimization of the design could yield substantial aggregated benefits. An important feature of the the three most successful sakia variants tested is that the outer compartments divided by the internal baffle plates discharge first into individual collection chambers which in turn discharge through holes surrounding the hub instead of having a common discharge orifice as on the more traditional designs. This prevents water running back into the compartment adjacent to it.

The types of sakia with separate discharge points for each compartment are distinguished by the generic name "tablia". A further advantage of the tablia type of device is that the water discharges a few centimeters above the centre shaft and therefore increases the useful head in relation to the diameter; especially with smaller machines. Typically a 3m tablia will lift water 1.5m compared with 0.90m for a centre-discharge sakia.

Another important conclusion from the tests by HRES was that for wheels operated in the 2-15rpm range, 6-8 compartments provide the optimum discharge. According to Molenaar [1], the following performance might be expected from traditional sakia designs:

diameter of sakia	head lifted	output
5m	1.8m	36m^3/h
4m	1.3m	51m^3/h
3m	0.9m	75m^3/h
2m	0.3m	114m^3/h

Comparison of the above outputs with those from a traditional persian wheel indicate that the sakia is somewhat more efficient, although of course it cannot lift water as high as is possible with a persian wheel.

3.5 RECIPROCATING DISPLACEMENT PUMPS

Water is for most practical purposes incompressible. Consequently, if a close fitting piston is drawn through a pipe full of water (Fig. 28 A), it will displace water along the pipe. Similarly, raising a piston in a submerged pipe will draw water up behind it to fill the vacuum which would otherwise occur (Fig. 28 B); this applies of course only up to a certain limit of the height water can be pulled by a vacuum, as discussed earlier in Section 2.1.5. In the first case water is displaced by the piston, but in the second case, the piston serves to create a vacuum and the water is actually displaced by atmospheric pressure pressing on its external surface, as indicated in the figure. So water can be displaced either by "pushing" or by "pulling", but it can also be "displaced" by a solid object being pushed into water so that the level around it rises when there is no where else for the water to go, as indicated in Fig. 28 C.

The displacement principle can be applied either through reciprocating/cyclic mechanisms, or continuously via rotary devices. The following sections deal first with reciprocating displacement pumps and later with rotary displacement pumps.

3.5.1 Piston or Bucket Pumps: Basic Principles

The most common and well-known form of displacement pump is the piston or "bucket" pump, a common example of which is illustrated in Fig. 29. These work by applying both the principles shown in Fig. 28 A and B; i.e., in the example of Fig. 29, water is sucked into the cylinder through a check valve on the up-stroke, and the piston valve is held closed by the weight of water above it (as in Fig. 28 B); simultaneously, the water above the piston is propelled out of the pump as in Fig. 28 A. On the down-stroke, the lower check valve is held closed by both its weight and water pressure, while the similar valve in the piston is forced open as the trapped water is displaced through the piston ready for the next up-stroke.

Fig. 30 shows a typical traditional design of brass-lined cylinder borehole pump with a metal foot valve and a metal piston valve; the piston has two leather cup-washer seals (indicated on the diagram). The outer casing and end fittings are normally cast iron in a pump of this kind.

There are various basic relationships between the output or discharge rate (Q), piston diameter (d), stroke or length of piston travel (s), number of strokes per minute (n), and the volumetric efficiency, which is the percentage of the swept volume that is actually pumped per stroke (η_{vol}):

if the swept area of the piston is $A = \dfrac{\pi d^2}{4}$

the swept volume per stroke will be $V = As$

the discharge per stroke will be $q = \eta_{vol} V$

the pumping rate (per minute) is $Q = nq$

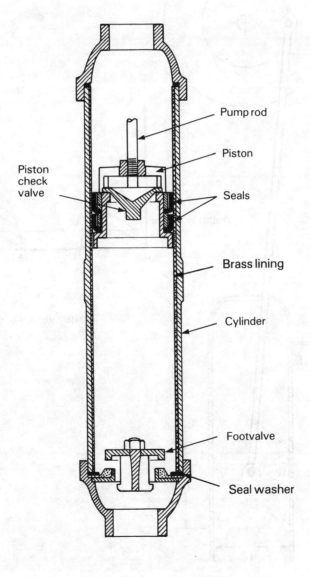

Fig. 30　Piston pump for use in borehole

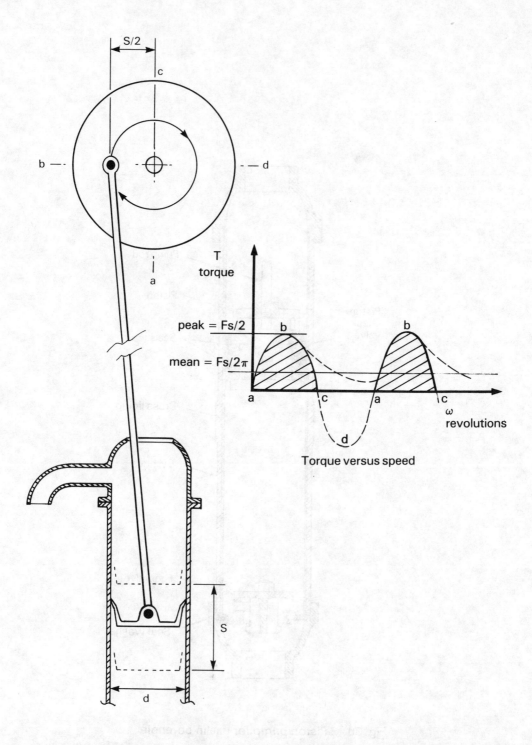

Fig. 31 Crank operated piston pump (valve details not indicated)

$$\text{hence } Q = 60n\eta_{vol}s\pi d^2/4$$

(multiplying by 60 gives Q in cubic metres per hour)

To use this result, if n is known in strokes/minute (or rpm), volumetric efficiency should be expressed as a decimal fraction (eg. 0.9), with d and s in metres. To convert the answer to litres per second simply divide the above answer by 3.6, (1,000 litres/m³ ÷ 3,600 seconds per hour).

Volumetric efficiency may be expressed as a decimal fraction and is sometimes called "Coefficient of Discharge". Another commonly used and related term is "Slippage" (X); this is the difference between the swept volume per stroke and the output per stroke; i.e.:

$$\text{slippage } X = V - q$$

Slippage arises partly because the valves take time to close, so they are often still open when the piston starts its upward travel, and also because of back leakage past piston or valve seats. Slippage is therefore normally less than unity, typically 0.1 or 0.2; it tends to be worse with shorter stroke pumps and with higher heads. With high flow rates at low heads, the moving water in the pipe can sometimes keep moving upwards with both valves remaining open for part of the down stroke so that discharge continues for part of the down stroke as well as on the up stroke. In such situations the "slippage" will be less than zero (known as "negative slip") i.e. the pump passes more water per stroke then its actual swept volume. In extreme cases with high speed pumping at low heads, the slippage can be in the region of -0.5, giving an equivalent volumetric efficiency in the region of 150%.

The force (F) required to lift the piston, will be the weight of the piston and pump rods (Wp), plus the weight of the column of water having a cross section equal to the piston area and a height equal to the head (H). There is also a dynamic load which is the force needed to accelerate these masses. If the acceleration is small, we can ignore the dynamic forces, but in many cases the dynamic forces can be large; dynamic loads due to accelerating water are discussed later in Section 3.5.4. In principle, the dynamic force, to be added to the static force, will be the summed product of the mass and acceleration of all moving components (i.e. water, plus piston, plus pump rod). In situations such as boreholes, where the pump rods are submerged within the rising main, their weight will be partially offset by an upward buoyancy force due to displacing water (see Fig. 38), which also should come into the equation. However this is not a text book on pump theory, so if we ignore possible "second order effects" such as buoyancy of pump rods and the dynamic forces, for simplicity, we arrive at a figure for the primary "static" force necessary to initiate movement of the piston;

$$F = W_p + A\rho gH$$

if W_p is in Newtons (or kilograms × 9.81), A is in m², ρ the density of water is 1,000 kg/m³, g is 9.81 m/s² and H is in metres, then F will be in Newtons.

If the pump rod is connected to a lever, as in a hand pump (Fig. 29), then the downward force required to lift the pump rod will be reduced by the ratio of the leverage, however, the distance the hand of the operator will have to move, compared with the stroke, will be proportionately increased.

The pump rod can also be connected to a flywheel via a crank (as in Figs. 31 and 89); this is the coventional way of mechanizing a reciprocating piston pump. The torque (or rotational couple) needed to make the crank or flywheel turn will vary depending on the position of the crank. When the piston is at the bottom of its travel (bottom dead centre or b.d.c.), marked as "a" on the figure, the torque will be zero as the pump rod pull is acting at right angles to the direction of movement of the crank and simply hangs on the crank; as it rotates to the horizontal position marked "b", the torque will increase sinusoidally to a maximum value of Fs/2 (force F times the leverage which is s/2); the resisting force will then decrease sinusoidally to zero at top dead centre (t.d.c.) marked "c". Beyond t.d.c. the weight of the pump rod and piston will actually help to pull the crank around and while the piston is moving down the water imposes no significant force on it other than friction. If, for convenience, we assume the weight of the piston and pump rod is more or less cancelled out by friction and dynamic effects, the torque is effectively zero for the half cycle from t.d.c. at "c" through "d" to "a" at b.d.c. where the cycle restarts. The small graph alongside the sketch in Fig. 31 illustrates the variation of torque with crank position through two complete revolutions; anyone who has turned a direct driven hand pump via a crank and hand-wheel will have experienced how the load builds up in this way for a quarter cycle and falls back to (near) zero for the next quarter cycle.

If the crank has a flywheel attached to it, as it normally will, then the momentum of the flywheel will smooth out these cyclic fluctuations by slowing down very slightly (too little to be noticeable) during the "a-b-c" part of the cycle and speeding up during the "c-d-a" part, as illustrated by the broken line following the first revolution in the graph in Fig. 31. If the flywheel is large, then it will smooth the fluctuations in cyclic torque to an almost steady level approximating to the mean value of the notched curve in the figure. The mean value of half a sine wave, to which this curve approximates, is the peak value divided by pi (π), (where $\pi = 3.142$).
Therefore:

$$\text{peak torque felt by a crank drive} = Fs/2$$
$$\text{mean torque (with flywheel)} \quad T = Fs/2\pi$$

Therefore, the torque necessary to turn a crank through its first revolution will be about π (i.e. approximately three) times greater than the mean torque which is needed to maintain steady running. Many prime movers cannot readily produce three times the torque needed for running in order to start a pump, and even with those that can, there is usually a price to pay to achieve this requirement. This is one reason why centrifugal pumps rather than piston pumps are more commonly used with engines and electric motors, as they actually need less torque to start them than to run them.

Power can be calculated as the product of speed and torque. Hence if the rotational speed is n (rpm) then a measure of power can be obtained as:

$$\text{power} \quad P = T(2\pi n/60) \text{ Watts}$$
$$= Fs(n/60) \text{ Watts}$$

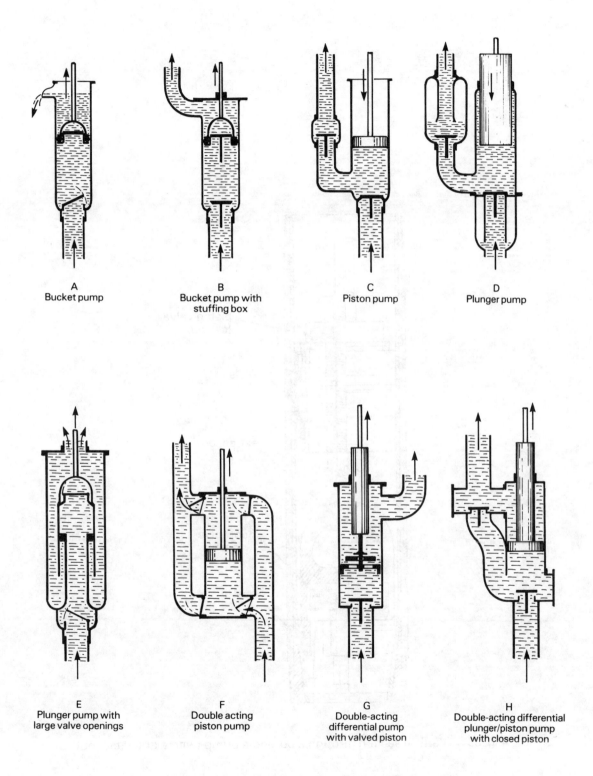

Fig. 32   Different types of reciprocating displacement pumps

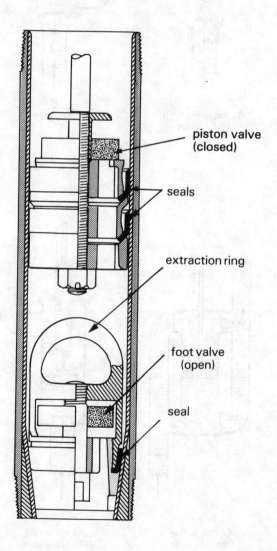

Fig. 33 Vertical section through a borehole pump (with extractable foot valve)

3.5.2 Double-acting Piston Pumps and Plunger Pumps

A single-acting pump only discharges water when the piston rises (if mounted vertically) and the down stroke is utilised simply to displace more water into the working space ready for the next stroke. It is possible to arrange things so that while one side of the piston displaces water to discharge it, the other induces more water, so that discharge takes place on both the up and the down stroke. Such pumps are known as "double-acting" pumps and are significantly more productive for their size than single-acting ones.

Fig. 32, diagrams A to H, illustrate various reciprocating displacement pump principles, and shows various single and double-acting configurations:

Diagram A in the figure shows a conventional single acting pitcher pump, as in Fig. 29.

Diagram B shows how, if the water needs to be delivered to a level higher than the point at which the pump rod enters the pump, a sealed "lid" is needed to prevent leakage. The seal is usually achieved by fitting a stuffing box through which the pump rod passes.

Diagram C is a piston pump in which the cylinder carries no valve; in principle it is similar to the pump in diagram B. It has the advantage that no stuffing box is needed, but it has the disadvantage that the discharge stroke requires the piston to be pushed rather than pulled, which needs a much stiffer pump rod in order to avoid buckling. There can also be problems with this kind of pump because the direction of motion of the water is reversed. This action can only be performed slowly, especially at low heads with large volumes of water per stroke, or sudden stopping and starting of the mass of the water will cause "water hammer" (much as when a tap or valve is suddenly closed and causes a "bang" in the pipes). Water hammer can damage or even burst a water system and is of course to be avoided. Pumps of this kind are therefore unusual today.

Diagram D is a similar pump to C, except it is a plunger pump rather than a piston pump. Here a solid plunger, sealed with a large diameter stuffing box or gland packing displaces the water; this is a more robust pump than C. The main justification for using plunger pumps is that the piston or plunger seals are less prone to wear through abrasive solids in the water, and also, where very high pressures and low flow rates are needed, a smaller plunger or closed piston is possible because a through valve is not needed through its centre. Therefore the main use today for pumps of this kind is for pumping small volumes of water up to very high pressures or heads, such as for reverse osmosis desalination plants, where pressures of the order of 300m of water are required. The plunger pump also suffers from the flow reversal problem of pump C, but this is less serious where small flow rates at high heads are involved. However the diagram shows a pump with air chambers below the inlet valve and above the delivery valve which are necessary to cushion the shocks caused by sudden reversal of flow direction, as explained in more detail in Section 3.5.4. Plunger pumps offered a manufacturing advantage in the past, in that it was sometimes easier to produce a good external finish on a plunger than inside a cylinder, but modern pump production techniques have reduced this advantage.

Diagram E indicates one of several methods to obtain large valve openings; this is important for low head pumps where high flow rates are required and it is necessary to minimise the hydraulic losses caused by forcing a lot of water through a small opening. Here the piston is external to a seal rather than internal to a cylinder; another way of looking at it is that a cylinder is being pulled up and down over a fixed piston.

Diagram F shows a pump that is similar in principle to that in C, but double-acting. Here when the piston is on the upstroke it induces water into the lower chamber and discharges from the upper, while on the downstroke water is induced into the upper chamber and discharged from the lower. The same points that apply to pump C apply to this design.

Pump G is known as a differential pump and is also double-acting; the pump rod is of a large diameter where it enters the upper chamber and if it is sized so that its cross-sectional area is exactly half the cross sectional area of the chamber, it will therefore displace half the volume of the chamber on the downstroke (the principle being as for plunger pump D), but on the up stroke, the other half of the volume will be discharged by the upward movement of the piston, as for bucket pump B.

Pump H applies a similar differential double-acting principle to pump G, but uses a closed piston as for pump C, and apart from being a more complicated arrangement, will be more prone to water hammer due to the flow reversal involved in both chambers.

Of the above configurations, only A, B, E and G are generally appropriate for irrigation pumping duties.

Although more complicated than single-acting pumps, double-acting pumps have considerably smoother outputs, and a smoother torque requirement. They were therefore widely used in conjunction with reciprocating steam engines, but the move to mechanize with electric motors or high-speed diesel engines has made them rarely used today, mainly because they have the following disadvantages;

i. they are larger and more complicated and hence more expensive

ii. they usually involve flow reversal which can cause waterhammer so some need to be run quite slowly or to incorporate air chambers

iii. the drive requires that the pump rod is pushed as well as pulled (at least with the configurations shown), so there must be no backlash or free-travel in the transmission (or hammering and wear and tear will result); also the pump rods must be capable of taking the compressive load on the down- stroke without buckling.

All this involves heavier more precisely engineered components in the drive train, which generally adds to the cost. Therefore, the trend in recent years has been to restrict the use of piston pumps to simple single-acting

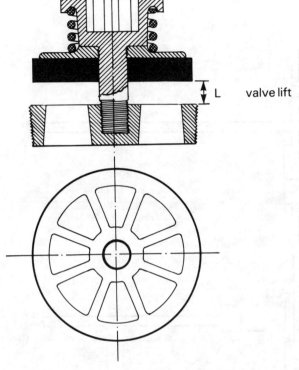

Fig. 34 Typical pump valve (shown open in cross-section)

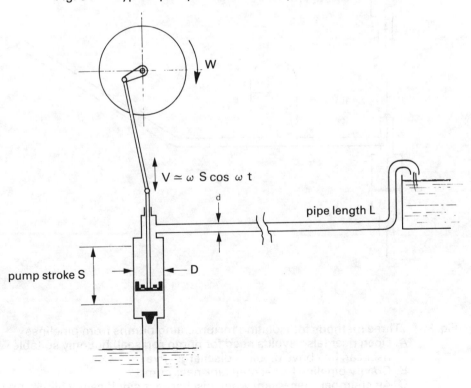

Fig. 35 Piston pump connected to a pipeline

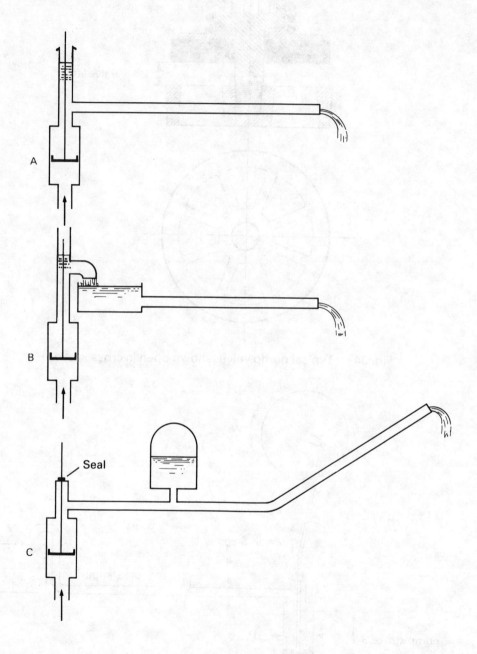

Fig. 36　Three methods for isolating reciprocating pumps from pipelines
A. Open riser (also avoids need for pump rod seal) but only suitable where riser can be above pipeline discharge level
B. Gravity pipeline from small tank near pump
C. Air chamber; necessary when discharge is significantly higher than pump

bucket pumps where their essential simplicity and low cost provide the justification for choosing them.

3.5.3 Pistons and Valves

Fig. 33 shows details of a typical borehole pump's piston and foot-valve. The simplest type of seal, commonly used in single-acting bucket pumps, is the leather cup washer as shown. Suitable grades of leather, commonly impregnated with "neatsfoot oil" boiled from the hooves of cattle, will function for surprisingly long periods (several years) in smooth drawn brass cylinders, or in smooth PVC. With the high cost of servicing deep boreholes, it is worth paying a premium to get a good life out of pump seals. Various synthetic leather "compound" materials based on plastics have been used for seals; these are often more consistent in their performance than leather and will often have better wear resistant characteristics.

Pistons can also be sealed by packings or piston rings. Packings generally need to be compressed by a certain optimum amount; too much compression, and friction and wear become excessive, while too little and there is excessive leakage around the piston. Graphited asbestos is the traditional industrial packing material but graphited PTFE (poly-tetrafluoroethylene) is now becoming available and offers superior sealing and wear characteristics. Similar packings are used around pump rod seals for reciprocating pumps and also sometimes for shaft seals on various rotary pumps. In all these applications they are generally compressed by screwing down a fitting against them in order to provide the sealing pressure; this needs regular adjustment.

All reciprocating pumps (and some rotary pumps) depend on check valves (sometimes known as non-return valves) which as their name suggests, allow water to flow one way but not the other. There are basically three categories of check valve:

i. Flexible valves that normally lie in a closed position, but open by being bent or deformed when pressure is applied

ii. Hinged valves that open like a door

iii. Straight lift valves which rise vertically and evenly from their seats

Fig. 34 shows a typical check valve design of the kind that may be used in a reciprocating pump. Valves are invariably opened by the difference in water pressure across them created by piston movement, but they may be closed again either by their own weight usually in combination with the weight of water trying to flow backwards. In some cases closing is assisted by a light spring, (as shown in the figure). Valve springs are usually made of bronze to avoid corrosion problems, but alternatively, valves may be made from an elastic material like rubber.

The main requirements of valves are a good seal when closed combined with lack of resistance to flow when they are open, and rapid opening and closing while achieving good durability. Usually rubber or alternatively precision ground metal mating surfaces are necessary to ensure there are no leakage gaps when the valve is closed. Effective sealing is particularly important with foot valves. To offer as little resistance to flow as possible when open (and to be capable of opening and closing quickly) demands large port areas with as few changes of flow direction as possible and sharp surfaces that can cause turbulence minimized. A rule of thumb sometimes used is that the suction valve should have a port area of at least two-thirds of the piston area, while the discharge valve (or piston valve) should have an area of at least half the piston area, [14]. Finally, rapid opening and closing (to minimise back-leakage) depends on light weight for the valve, combined with a short travel; light weight can demand a trade-off with robustness and durability, while short travel conflicts to some extent with the need for a good unobstructed passage for water when the valve is open. Therefore all valves are a compromise in achieving conflicting requirements.

Finally, valves are the main mechanical components of a pump and so are subject to wear and tear. It is therefore desirable to use pumps in which the valves and their seats can readily (and inexpensively) be replaced when necessary.

3.5.4 Reciprocating Pumps and Pipelines

A reciprocating pump moves water in a non-continuous manner, so the water is constantly accelerated and decelerated by the movement of the piston. Very large forces can be created if long pipelines containing a large mass of water are directly connected to a reciprocating pump. This is because the pump piston tries to force the water in the pipeline to move rapidly from rest to speed, and then back to rest; since water is incompressible it will try and follow the motion of the piston. Therefore reciprocating pumps need to be isolated from water in long pipelines by methods described shortly, in order to cushion the water in the pipeline from the motion of the piston.

To gain an appreciation of the damage that can happen and the consequent importance of isolating reciprocating pumps from pipelines, it is worth running through some simple calculations to quantify the forces concerned. Fig. 35 illustrates a simple piston pump (diameter D by stroke S) with a long length (L) of delivery pipeline (diameter d). Newton's Laws of Motion, state that a force is necessary to accelerate a mass from one velocity to another (or from rest); this force is numerically equal to the product of the mass and the acceleration at any moment in time. If we assume the pump piston is driven sinusoidally, (as it would be if driven by a steadily revolving crank, having a long connecting rod in relation to the stroke), then the maximum acceleration of the piston (and hence of water being propelled by it) will be A_{max}, where:

$$A_{max} = -\omega^2(S/2)$$

where ω is the angular velocity of the driving crank in radians/second; (2.radians = 1 revolution or 360°).

The acceleration of the water in the pump will be magnified for the water in the pipe, since if the pipe cross sectional area is smaller than that of the pump, a higher velocity will be needed to pass the same flow of water, and hence a proportionately higher acceleration to reach the higher velocity. The magnification will be proportional to the ratio of pump cross sectional area to pipe cross sectional area, which in turn is proportional to the ratio of their diameters squared; hence the acceleration of water in the pipeline will be:

$$A_{pipe} = (D/d)^2 A_{piston}$$

The force necessary to achieve this acceleration for the water in the pipeline will be equal to the mass of water flowing multiplied by its acceleration. The mass is the volume of water in the pipe times its density; hence the accelerating force is:

$$F = (\pi d^2/4) - L \rho (D/d)^2 A_{piston}$$

the maximum force will occur at the moment of maximum acceleration. So, assuming sinusoidal motion of the piston and the flow, this is numerically such that:

$$A_{piston}(max) = -\omega^2(S/2)$$

Hence, if m is the mass of water in the pipeline:

$$F_{max} = -m(D/d)^2 \omega^2 (S/2)$$
$$F_{max} = -(\pi d^2/4) L \rho (D/d)^2 \omega^2 (S/2)$$

For example: suppose we consider a pump of 100mm diameter, connected to a pipeline of 50mm diameter, density of water 1,000kg/m³, a crank speed of 60rpm (which is 1 rev/s or (1 x 2)rad/s) and a stroke of 300mm (0.3m), then:

$$F_{max} = -(\pi/4)(0.05)^2 L \cdot 1000(0.1/0.05)^2 (2\pi)^2 (0.3/2)$$

$$F_{max} = -46.5\,L \quad \text{Newtons}$$

In other words, there is a peak reaction force due to the water in the pipe of 46.5N per metre of pipe, due to sinusoidal acceleration under the conditions described. A 100m pipeline will therefore, with conditions as specified, experience peak forces of 4,650N while a 1 km pipeline will experience 46,500N - and so will the pump in both cases; (46,500N is the equivalent of about 4.5 tonne, or 10,000 lb).

Since this force is proportional to the square of the pump speed, doubling the pump speed to 120 strokes per minute will impose four times the acceleration and hence four times the force. Even such modest pump speeds (by rotary pump standards) as say a few hundred rpm will therefore impose impossible accelerative forces on the water in the pipe line <u>unless it is isolated or cushioned from the motion of the pump piston.</u>

In reality, the situation is not quite as bad, as even steel pipes are flexible and will expand slightly to take the shocks. But in some respects it can also be worse, because when valves slam shut, very brief but large shock accelerations can be applied to the water; these are known as "water hammer" because of the hammering noise when this happens. Water hammer shocks can damage both a pump, and its prime mover, as well as possibly causing burst pipes or other problems.

The same problem can occur on suction lines as well as on delivery lines, except that a sudden drop of pressure caused by high flow velocities can cause "cavitation" where bubbles of water vapour and dissolved air suddenly form. When the pressure increases again slightly, the bubbles can then suddenly catch up and the bubble will implode violently causing water hammer.

When the pump outlet is set close to or above the pipeline discharge level, there is no great problem because the pipeline can be de-coupled from the pump by feeding into a small tank which can then gravity feed the pipeline steadily; see Fig. 36 B. Alternatively, a riser open to the atmosphere in the pipeline near to or over the pump can achieve the same effect (as in Fig. 36 A); because the pump rod can go down the riser it neatly avoids the need for a seal or stuffing box.

Where the pump delivers into a pipeline which discharges at a significantly higher level, it is generally not practical to have a riser open to the atmosphere at or near the pump, since it obviously would have to extend to a height above the level of the discharge. The solution generally applied in all such cases where more then a few metres of suction or delivery line are connected to a reciprocating pump is to place an air chamber or other form of hydraulic shock absorber between the pump and the pipeline (Fig 36 C) and always as close to the pump as possible to minimize the mass of water that is forced to follow the accelerations of the piston. Then when water from the pump seeks to travel faster than the water in the pipeline it will by preference flow into the air chamber and compress the air inside it. When the piston slows so that the water in the pipeline is travelling faster than that from the pump, the extra water can flow out of the airchamber due to a slight drop in pressure in the pipeline and "fill the gap". In otherwords, an air chamber serves to smooth the flow by absorbing "peaks" in a reciprocating output and then filling the "troughs" that follow the peaks.

Air chambers are generally vital on long or on large capacity pipelines when using a reciprocating pump (eg. Fig. 106), but they are well worth their extra cost not only in reducing wear and tear, but also the peak velocity of water in the pipeline will be reduced which in turn reduces pipe friction; this reduces the power requirement and saves pumping energy.

A special problem with air chambers on delivery lines is that the air in the chamber can gradually dissolve in the water and be carried away, until there is no air left and water hammer then occurs. Therefore simple air chambers usually require regular draining to replenish their air by opening a drain plug and an air bleed screw simultaneously, obviously when pumping is not taking place. Suction line air chambers are usually replenished by air coming out of solution from the water, although when air-free groundwater is being drawn, a small air snifting valve may be needed to deliberately leak in a minute flow of air and prevent the chamber losing its air volume. Industrial air chambers sometimes contain a sealed rubber bag which will retain its air indefinitely; these are recommended in situations where regular attention cannot be guaranteed. Spring loaded hydraulic shock absorbers have also been used in the past instead of air chambers (Fig. 37). Another useful

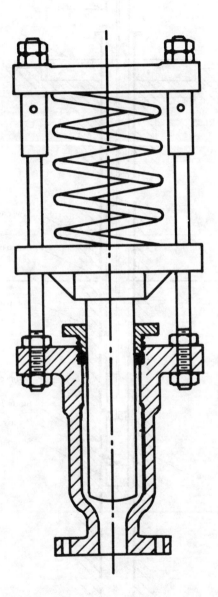

Fig. 37   Hydraulic shock absorber can serve as an alternative to an air chamber, especially with very high pressure pipelines

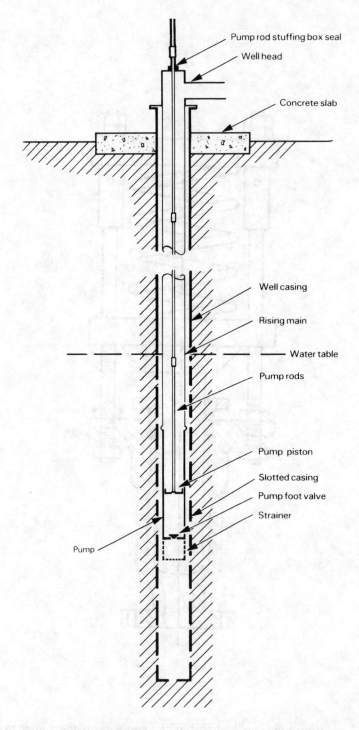

Fig. 38   Schematic cross-section through a borehole

alternative is to pump into a rubber or other type of flexible pipe which is less prone to water hammer than a steel pipeline. Care must be taken with rigid plastic pipelines as they can easily be broken by water hammer, especially in cold weather when they tend to be more brittle.

A typical size for an air chamber will be around twice the swept volume of the pump, however it will need to be larger to cater for more severe flow irregularities or long delivery lines generally.

3.5.5 Reciprocating Borehole Pumps

When groundwater is deep, or the ground is hard to dig, boreholes are generally quicker and less expensive to construct than dug wells. Most boreholes are lined with either a 100mm or 150mm (4in or 6in) steel rising main, so the pump must be small enough in diameter to fit down it. Fig. 38 illustrates a typical borehole with a piston pump in it.

Fig. 33 shows a typical borehole piston pump; because bore hole diameters need to be kept small (large boreholes are very expensive to drill) the best way to obtain a larger output from a borehole pump is to increase its stroke. The speed of operation of borehole pumps is usually restricted to about 30 strokes a minute, although a few operate at up to 50 strokes per minute. Higher pumping speeds tend to buckle the long train of pump rods by not giving the pump rods and piston sufficient time to fall back on the down stroke under gravity. Obviously, with any depth of borehole, it is important to avoid compressing the pump rods, or they may buckle and jam against the sides of the rising main.

The normal method of installing or removing a pump from a borehole is to raise the entire rising main using block and tackle or a crane, section by section. The procedure is to lift a complete section clear of the well; clamp the next section below it and lower the whole assembly so it hangs on the clamp. The top 6m (20ft) section standing clear of the well can then be unscrewed and the process is then repeated. Since this is a long and expensive process, it is important to minimize the number of times it is necessary to do this in the life of a borehole. It is common therefore to install pumps with a piston diameter slightly smaller than the diameter of the rising main, especially on deep boreholes. This allows the piston to be pulled up through the rising main by lifting only the pump rods. In fact, Fig. 33 shows an extractable borehole pump of this kind; this pumps is designed so that the footvalve can also be removed without removing the cylinder and rising main. One method, as shown in the figure, is where a loop on top of the footvalve can be fished for, caught, and used to pull it out. Alternatively, with some other types of footvalves, the pump rod is disconnected from the drive at the surface, and the piston can then be lowered onto the footvalve, which has a threaded spigot on top of it which will screw into the base of the piston if the pump rods are twisted. Then when the pump rods are raised, both footvalve and piston can be drawn up to the surface together. The importance of this is that piston and footvalve seals are the primary wearing parts needing regular replacement. Therefore, the ease with which a borehole pump can be overhauled is an important consideration for all but the shallowest of boreholes.

Pump rods can be improvised from galvanized water pipe joined with standard pipe sockets, but properly purpose-designed pump rods are commercially available, and although more expensive, are more easily connected or disconnected and are also less likely to break or to come unscrewed. A broken pump rod at best is difficult to recover and at worst can mean the loss of an expensive borehole. Where wood of suitable (very high) quality is available, plus appropriate couplings, wooden pump rods have the advantage that they do not impose any extra weight, which hinders initial start up of a reciprocating pump in a borehole, because they float in the water of the rising main.

The main attraction of using reciprocating pumps driven by tensile pump rods in boreholes is that they are essentially simple and the better commercial products have become highly reliable in operation. Typically, pump rods in tension can pull well over a tonne, allowing the use of this type of system on boreholes as deep as 300m (1000ft); this is the standard system used for farm windpumps and quite a few of the more traditional diesel powered systems use a gearbox and pitman mechanism to drive a reciprocating borehole pump in exactly the same way.

A common problem with boreholes is that they are often not truly vertical and sometimes they are curved, making any pump rods scrape the rising main and eventually wear a hole in it. Sometimes boreholes start off straight and earth movements cause subsequent distortion of the bore. Also, removing a pump to change its seals, with its long train of pumprods is a slow and expensive operation. Therefore there has been an incentive to find other methods of driving borehole pumps than by pump rods.

3.5.6 Hydraulically activated Borehole Pumps

An alternative method for powering reciprocating borehole pumps to pump rods, is with an hydraulic transmission. Here water under pressure is used to push more water to the surface.

An example is the Vergnet "Hydro-pompe", illustrated in Fig. 39, which has been quite successfully used for small water supplies, particularly in the West African Sahel region, but the output is probably on the low side for most irrigation applications. It works by a foot pedal which is mounted on a pilot piston to force water down a flexible pipe. The pump cylinder, which is located below water level in the well or borehole, has a conventional "suction valve" or foot valve and strainer, a pump chamber, and at the top of the chamber a discharge check valve with a discharge pipe leading up to the surface above it. Instead of a piston to displace water in the cylinder there is what Vergnet call a "Diaphragmatic Hose", this is in fact an elastic bladder which expands to displace water from the pump chamber when the foot pedal is depressed and pumps water into it. A similar hydraulic transmission, called the "Hidromite" system, for driving a reciprocating borehole pump, was developed in Australia mainly for use with windmills. Here the windmill, (or other prime-mover), drives a master piston located at the surface. This is a double-acting water pump connected by two hydraulic transmission pipes to a slave piston at the bottom of the borehole, directly connected by a short pump rod to the actual pump piston below it, which operates in the conventional way.

Although hydraulic transmission units are quite attractive in some respects compared with pump rods, they are significantly more complicated and expensive. Their efficiency is also likely to be lower, due to the extra pipe friction involved in moving the water needed to power the submerged pump.

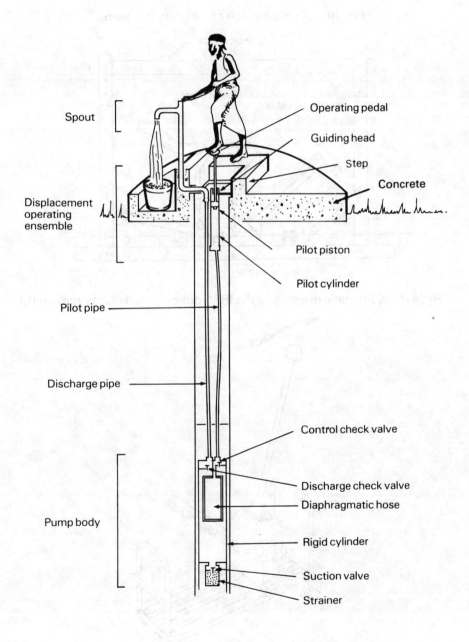

Fig. 39 The Vergnet hydraulic foot-pump

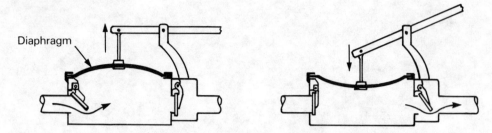

Fig. 40   Cross-section of a diaphragm pump

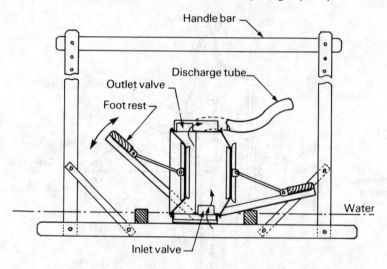

Fig. 41   Schematic drawing of the IRRI foot-operated diaphragm pump

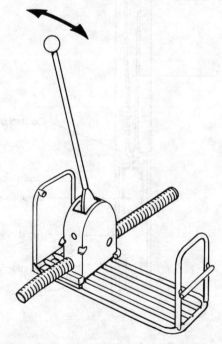

Fig. 42   Commercial portable double acting diaphragm pump

3.5.7 Diaphragm pumps

An alternative to the use of a piston in a cylinder for pumping is to fit one wall of a pump chamber with a flexible diaphragm which when moved in and out displaces water (see Fig. 40). Here the left hand valve is analagous to the foot valve of a piston pump and the right hand one is the delivery valve.

In general, the advantages of a diaphragm pump are:

i. perfect sealing (except for any shortcomings of the two check valves);

ii. high mechanical efficiency, since flexing a diaphragm involves much less friction then sliding a piston with seals up and down a cylinder;

iii. no seal is needed at the pump rod which also reduces friction losses still further compared with piston pumps;

iv. they are self-priming, hold their prime very well and can often handle a higher than average suction head;

v. they often function well with gritty or muddy water which could damage a piston pump.

There are however also disadvantages:

i. diaphragms need to be high quality rubber if they are to last, and are therefore expensive;

ii. diaphragm pumps are often dependent on specialized spare parts that cannot easily be improvised in the field;

iii. a diaphragm pump is similar to a large diameter piston pump with a short stroke; so the pump rod forces are high in in relation to the head and swept volume. This imposes a high load on transmission components and on the point of attachment of the pump rod to the diaphragm;

iv. therefore diaphragm pumps (of the kind in Fig. 40) are only suitable for low head pumping in the 5-10m range.

Fig. 41 shows a foot-operated, double-acting diaphragm pump developed by the International Rice Research Institute (IRRI) in the Philippines for irrigation purposes. Unlike traditional devices such as Dhones or Shadoofs, this pump is portable (by two men) and can therefore be moved along an irrigation canal in order to flood one paddy after the other. However it is less efficient than the better traditional water lifters.

Fig. 42 shows a commercially manufactured, double-acting diaphragm pump that is mostly used for purposes such as dewatering building sites; it has the advantage of being portable, reasonably efficient and well suited to low heads and can deliver quite high outputs, so it, or similar designs, could equally be used for irrigating small landholdings. A pump of this kind was well liked by Ethiopian farmers irrigating small plots from the Omo River in a training project in which the author was involved [15].

One type of diaphragm pump that can be improvized and which reportedly works reasonably well at low heads is a design based on the use of an old car tyre as the flexible member (Fig. 43). Worn car tyres are of course widely available. The principle of this pump is to make a chamber by fitting end-plates into the openings of the tyre so that one is anchored and the other can be forced up and down. If suitable check valves are provided, this can make an adequate diaphragm pump. The prospective user should not underestimate the constructional requirements to make an adequately reliable device of this kind. For example a typical car tyre of 400mm overall diameter will have an effective area of $0.126m^2$; this requires a force of 1,230N per m head; i.e. only 3m lift requires nearly 3,700N pull to displace any water (this is the equivalent of 376kg force or 830 lb). Robust fixings and connections are therefore needed to prevent such a pump coming apart, even at quite low heads. One further problem reported with car tyre pumps is that they do not work well as suction pumps because the internal structure of a tyre can separate from the outer rubber casing when repeatedly pulled by suction pressures. Tyre pumps could make a useful high-volume low-head pump however, providing they are skillfully constructed, to be powered perhaps by two people working a suitably strong lever, and providing they operate submerged or with limited suction lift.

3.5.8 Semi-rotary Pumps

A form of reciprocating positive displacement pump, using the same principles as a piston pump, is the semi-rotary pump. Here a pivoted plate, or "bucket vane" can be reciprocated, like a door on hinges, through about 270° within a circular chamber. It alternately draws water from one side and then the other through check valves. The semi-rotary pump is mostly used as a hand-pump, often for pumping kerosene and fuel oils rather than water, and has only a small capacity. It is also sensitive to any dirt in the water, which can easily jam it. It is therefore unlikely to be useful for irrigation purposes and is simply mentioned for completeness.

3.5.9 Gas Displacement Pumps

Water can be displaced by a gas or vapour as readily as by a solid. A number of air and vapour displacement pumps were manufactured at the beginning of this century. The former rely on air delivered by an engine-driven compressor, while the latter generally used steam to displace water directly, rather than through the intermediary of a steam engine and pump. The Humphrey Pump is an analagous device which uses the gases generated in an internal combustion engine cycle to displace water directly in much the same way. Both compressed air and steam displacement pumps suffer from being inherently inefficient, as well as being massive (and hence expensive) in relation to their pumping capacity, but in contrast the Humphrey Pump is actually more efficient than most comparably sized conventional i.c. engine pumping systems, although it is also quite large.

Fig. 44 shows the principle of the Humphrey Pump, which consists in

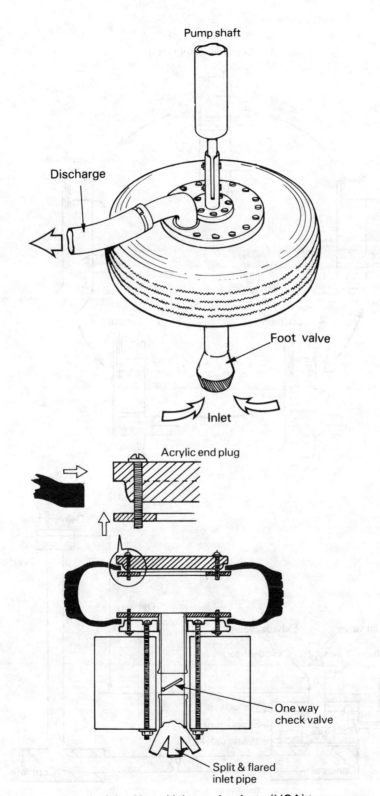

Fig. 43　Detail of the New Alchemy Institute (USA) tyre pump
Principle of operation of a 4-stroke Humphrey pump

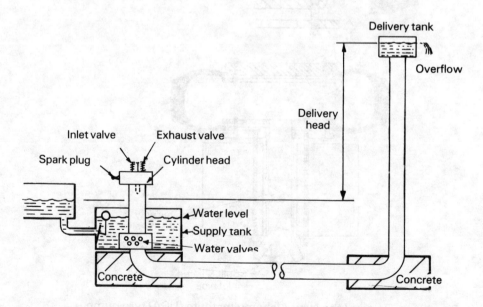

Fig. 44 The Humphrey pump: a liquid piston internal combustion engine and pump combined

effect of a conventional four-stroke i.c. engine cylinder head mounted on top of a pipe which forms the working space. This is in turn connected to a long horizontal pipe and a riser to the discharge level. The working space pipe is located over the water source and there are inlet valves to allow water to enter the system below. The Humphrey Pump cycle is similar to a standard four-stroke piston engine except that instead of the engine having a metal piston driving a crank shaft, the water in the working space acts as a piston. All the Humphrey Pumps so far built could only run on gaseous fuels such as coal gas or natural gas because of difficulties with vaporising liquid petroleum fuels successfully under the cool cylinder conditions which occur when cold water acts as a piston.

The Humphrey Pump's cycle is sequenced by a pressure sensor which controls a simple linkage to open and close the exhaust and inlet valves on the cylinder head at the correct times. The water in the long horizontal "U" pipe and riser oscillates, to provide the induction and compression strokes, before being driven forcibly along the pipe by the firing stroke. A vacuum created behind the departing column of water causes more water to be induced through the inlet valves.

Humphrey Pumps were used for irrigation projects in the USA, and Australia in the early part of this century, with some success, and the University of Reading in England has developed a modern small scale prototype intended for irrigation pumping with biomass fuels.

The main advantage of the Humphrey Pump other than good fuel efficiency, is its great mechanical simplicity. It therefore can readily handle muddy or sandy water and has the potential for extreme reliability, yet requires very little maintenance. The main negative features are the need for gaseous fuels and it can only readily operate from water sources where the water level does not change much.

3.6   ROTARY POSITIVE DISPLACEMENT PUMPS

There is a group of devices which utilize the displacement principle for lifting or moving water, but which achieve this by using a rotating form of displacer. These generally produce a continuous, or sometimes a slightly pulsed, water output. The main advantage of rotary devices is that they lend themselves readily to mechanization and to high speed operation. The faster a device can be operated the larger the output in relation to its size and the better its productivity and cost-effectiveness. Also, steady drive conditions tend to avoid some of the problems of water hammer and cavitation that can affect reciprocating devices.

Centrifugal pumps, which use a different principle and are described later, have in fact become the most general mechanized form of pump precisely because they can be directly driven from internal combustion engines or from electric motors. But rotary positive displacement pumps have unique advantages over centrifugal pumps in certain specialized situations, particularly in being able to operate with a much wider range of speeds or heads.

Some types of rotary positive displacement pump have their origins among the earliest forms of technology (eg.the Archimedean Screw), and even today lend themselves to local improvization. In the past, industrially manufactured rotary pumps were less successful than centrifugal ones, possibly because they suffered from a number of constructional and materials problems. But modern, tougher and more durable plastics and synthetic rubbers may well be an important factor in encouraging the manufacture of a number of new types of rotary positive displacement pumps which could be advantageous in some situations, as will be described.

3.6.1 Flexible Vane Pumps

Here a flexible toothed rotor is used, generally made of rubber, Fig. 45. This is very simple in concept, being like a revolving door, but it can involve both considerable friction and significant back leakage. It cannot therefore be considered as an efficient type of pump. On the positive side, it will readily self-prime and can achieve a high head at low rotational speeds. Much will depend on the quality of the rotor material and the type of internal surface of the casing so far as both friction and durability are concerned.

Another similar type, developed recently by Permaprop Pumpen in Germany, has an endless rubber toothed belt which is driven around two pulleys; (see Fig. 46). As it curves around a pulley, the teeth on the belt spread apart and increase the volume between them, thereby drawing in water. The diagram shows how both sides of the chamber simultaneously pump in opposite directions, and suitable channels in the casing direct the water. The advantages claimed by the manufacturers are, _inter alia_, that it can run on "snore" indefinitely - (i.e. pumping a mixture of air and water), it will readily self-prime and suck water up to 8m and lift it a further 45m under the power of a small portable single cylinder engine. It is therefore a much more versatile pump than the equivalent centrifugal pump, but it is more complicated and expensive.

3.6.2 Progressive Cavity (Mono) Pumps

None of the rotary pumps so far reviewed lend themselves to being lowered down boreholes; in fact their main selling point is as suction pumps. However, the "progressive cavity" alias "progressing cavity" alias "Mono" pump (after its French inventor, Moineaux), (see Fig. 47) is unique in being a commercially available rotary positive displacement pump that readily fits down boreholes. This is a great advantage because positive displacement pumps can cope much more readily with variations in pumping head than centrifugal pumps. Therefore, any situation where the level may change significantly with the seasons or due to drawdown, or even where the drawdown is uncertain or unknown, makes the progressive cavity pump an attractive option. It also has a reputation for reliability, particularly with corrosive or abrasive impurities in the water. The reasons for this relate to good construction materials combined with a mechanically simple mode of operation.

Fig. 47 shows that this pump consists of just a single-helix rotor inserted in a double-helix stator. A single helix is rather like a simple spiral staircase while a double means two intertwined helixes. The stator helix is usually made from chromium plated steel or from stainless steel with a polished surface finish, and is circular in cross section and fits accurately into one of the two helices of the stator. The stator is usually

Fig. 45 A flexible vane pump

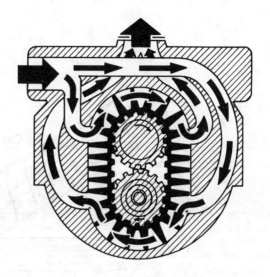

Fig. 46 The Permaprop tooth pump

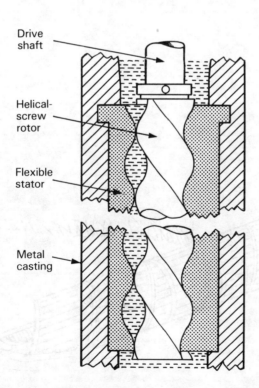

Fig. 47 Progressive cavity or 'Mono' pump

Schematic cross-section to illustrate principle of Archimedean screw

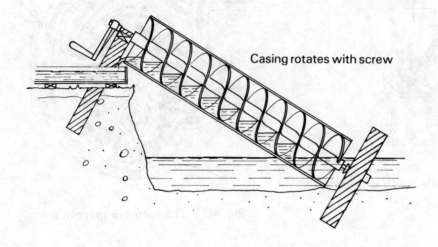

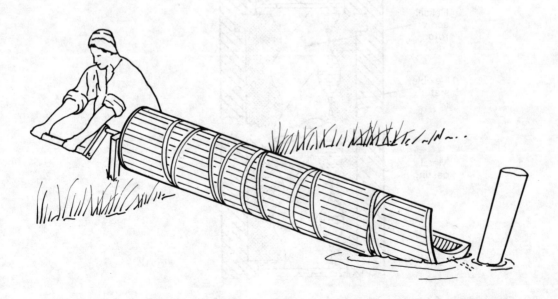

Fig. 48 An Archimedean screw. Two men are needed if the water head is more than 0.6 metres (See also Fig. 97 for an animal-powered version) (after Schioler [24])

moulded from rubber or plastic and the cross-section of its internal helix is oval, similar to two circles similar to the rotor abutting each other. A feature of the geometry of this type of pump is that the empty second start of the stator is divided into a number of separated empty voids, blocked from each other by the solid single start rotor. When the rotor is turned, these voids are screwed along the axis of rotation, so that when the assembly is submerged, discrete volumes of water will be trapped between the single start rotor helix and the inside of the double start stator in the voids and when the shaft is rotated these volumes of water are pushed upwards and discharged into a rising main.

Pumps of this kind are usually driven at speeds of typically 1000 rpm or more, and when installed down a borehole they require a long drive shaft which is guided in the rising main by water lubricated "spider bearings" usually made of rubber. Although friction forces exist between the rotor and stator, they are reduced by the lubricating effect of the flow of water, and they act at a small radius so that they do not cause much loss of efficiency. Progressive cavity pumps therefore have been shown to be comparably efficient to multi-stage centrifugal pumps and reciprocating positive displacement pumps under appropriate operating conditions. Their main disadvantage is their need for specialized components which cannot be improvised and their quite high cost; however, high cost is unfortunately a feature common to all types of good quality borehole pump and is usually justified by the need to minimise the frequency of the expensive procedure of removing and overhauling any pump from a deep borehole.

The progressive cavity pump can be "sticky" to start - i.e. it needs more starting torque than running torque (similarly to a piston pump) to unstick the rotor from the stator and get the water that lubricates the rotor flowing. This can cause start-up problems if electric motors or engines are used, but certain improved versions of this kind of pump include features which reduce or overcome this problem.

3.6.3 Archimedean Screw and Open Screw Pumps

The progressive cavity pump is one of the more recent pump concepts to appear, while the Archimedean screw is one of the oldest, yet they have a number of similarities.

Fig. 48 illustrates a typical Archimedean screw pump (and an animal-powered version is shown in Fig. 97). The traditional version of this pump, built since before Roman times and still used in a similar form in Egypt, is made up of a helix of square cross-section wooden strips threaded onto a metal shaft and encased in a tube of wooden staves, bound like a barrel with metal bands.

The Archimedean screw can only operate through low heads, since it is mounted with its axis inclined so its lower end picks up water from the water source and the upper end discharges into a channel. Each design has an optimum angle of inclination, usually in the region of 30°1to 40°, depending on the pitch and the diameter of the internal helix.

The principle is that water is picked up by the submerged end of the helix each time it dips below the surface, and as it rotates a pool of water gets trapped in the enclosed space between the casing and the lower part of each turn. As the whole assembly rotates, so the helix itself screws each trapped pool of water smoothly further up the casing until it discharges from the opening at its top; the water pools move much as a nut will screw itself up a bolt when prevented from rotating with it. This is also analagous to the trapped volumes of water screwed between the rotor and stator of progressive cavity pumps.

Traditional wooden Archimedean screws of the kind just described have been tested and found to have efficencies in the region of 30%.

The modern version of the Archimedean screw is the screw pump, Fig. 49. This consists of a helical steel screw welded around a steel tubular shaft, however unlike an Archimedean screw, there is no casing fixed to the screw, but it is mounted instead in a close fitting, but not quite touching, semi-circular cross-section inclined channel. The channel is usually formed accurately in screeded concrete. Because of the clearance between the screw and its channel, some back-leakage is inevitable, but the total flow rate produced by a screw pump is so large that the backflow is but a small percentage. Therefore modern screw pumps can achieve high efficiencies in the region of 60-70%.

Their primary advantage is that the installation and civil workings are relatively simple, compared with those for large axial flow pumps necessary to produce the same volume of output (which would need a concrete sump and elaborate large diameter pipework as in Fig. 66). Also, the screw can easily handle muddy or sandy water and any floating debris, which is readily pulled up with the water.

Probably the main disadvantage of screw pumps is that an elaborate transmission system is needed to gear down an electric motor or diesel engine drive unit from typically 1500 rpm to the 20-40 rpm which is normally needed. Mechanical transmissions for such a large reduction in speed are expensive and tend to be no more than 60-70% efficient, thereby reducing the total efficiency of the screw pump, including its transmission to about 50-60%. Screwpumps also present a safety hazard by having a potentially dangerous open rotor and should therefore be fitted with mesh guards. Finally, they cannot cater for much change in level of the water source, unless provision is made to raise or lower the entire unit and also the maximum head that can be handled will not exceed much more than 6m in most cases, and will normally be no more than 4 or 5m for smaller screw pumps.

3.6.4 Coil and Spiral Pumps

These pumps use a similar principle to the Archimedean screw except that they run horizontally while the Archimedean screw is tilted at about 30°. The coil and spiral pump family, if fitted with a suitable rotating seal, can deliver water to a greater height, typically 5-10m, above their discharge opening. Fig 50 A shows a spiral pump and B shows a coil pump.

Both these pumps work on the same principle, involving either a spiral or a coiled passage (in the latter case a coiled hosepipe serves the purpose) rotating on a horizontal axis. One end of the passage is open at the periphery and dips into the water once per revolution, scooping up a pool of water each time. Due to the shape of the spiral or the coil, sufficient water is picked up to fill completely the lower part of one turn, thereby trapping air in the

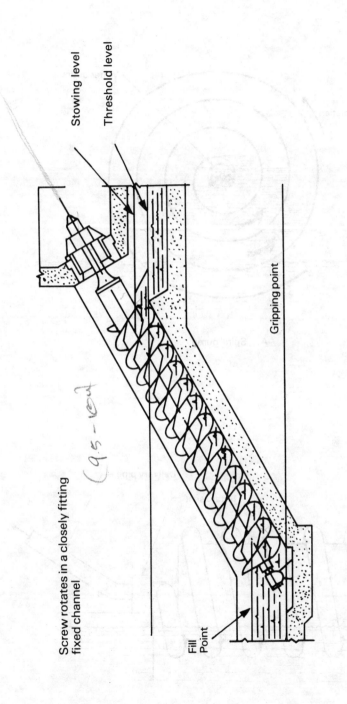

Fig. 49 Cross-section through an open screw pump

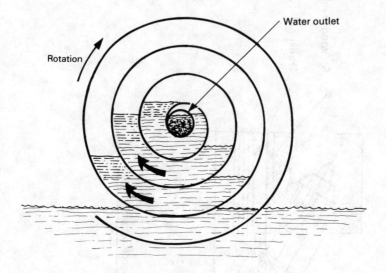

A. Spiral pump

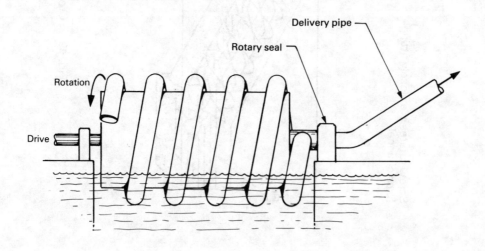

B. Coil pump

Fig. 50 Hydrostatic pressure pumps

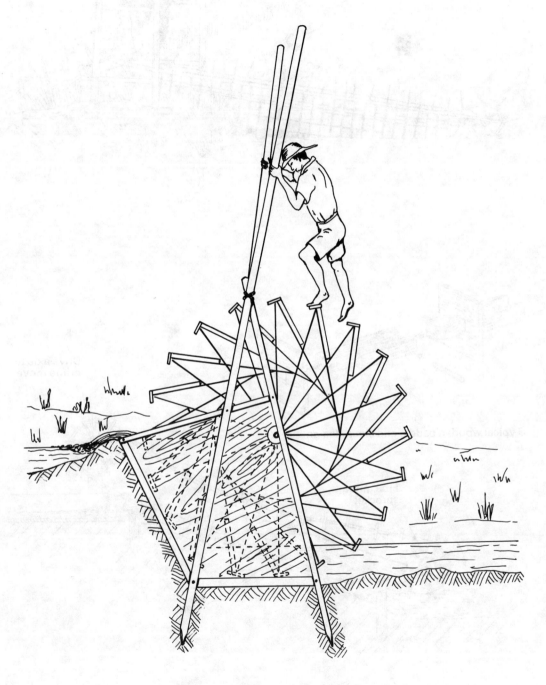

Fig. 51　Paddle-wheel or tread-wheel

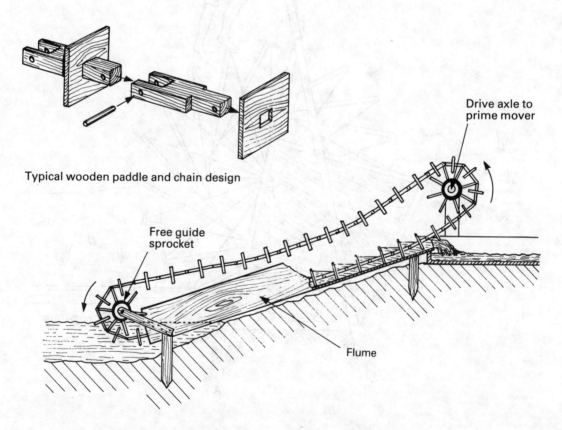

Fig. 52   Water ladder or Chinese 'Dragon Spine' pump

next turn. The pools of water progressively move along the base of the coil or of the spirals as the pump turns, exactly like an Archimedean Screw. However, when acting against a positive head, the back-pressure forces the pools of water slightly back from the lowest position in each coil as they get nearer to the discharge; so they progressively take up positions further around the coil from the lowest point. The maximum discharge head of either type of pump is governed by the need to avoid water near the discharge from being forced back over the top of a coil by the back pressure, so this is still a low head device.

The spiral pump has to be designed so that the smaller circumferences of the inner loops are compensated for by an increased radial cross section, so it would normally be fabricated from sheet metal; the coil pump is of course much easier to build.

This type of pump was originally described in the literature as long ago as 1806 and has attracted much fresh interest recently, with research projects on it at the universities of California (USA), Salford (UK), Los Andes (Colombia) and Dar es Salaam (Tanzania), [16]. Although historically the coil pump was used as a ship's bilge pump, today it is finding favour for use in river current powered irrigation pumps by, for example, the Royal Irrigation Department of Thailand (see Fig. 153) and also similarly by Sydfynsgruppen and the Danish Boy Scouts for an irrigation project in south Sudan, with support from Danida and on the Niger near Bamako in Mali, under a project supported by the German aid agency BORDA. Chapter 4.9 deals with some of the practical applications of this device.

The advantages of these devices are their inherent mechanical simplicity combined with the fact that, unlike an Archimedean screw, they can deliver into a pipe to a head of up to about 8-10 m, making them more versatile. The only difficult mechanical component is a rotary seal to join a fixed delivery pipe to the rotating output from the coil. They are ideal for water wheel applications due to the low speed and high torque needed, (which is where most of the research effort appears to be concentrated).

Their main disadvantage is that their output is small unless rather large diameter hose is used,, being proportional to the capacity of the lower part of one turn of hose per revolution. A simple calculation indicates that a significant and not inexpensive length of hose is needed to produce an adequate coil pump, (e.g. just 20 coils of only 1.5m diameter needs nearly 100mm of hose). Supporters of this concept argue that its simplicity, suitablility for local improvization and reliability should compensate for these high costs, but this type of pump has so far not been popularized successfully for general use and it does not exist as a commercial product.

3.6.5 **Paddle-wheels, Treadmills and Flash-wheels**

These devices are, in effect, rotary versions of the simple scoop; however instead of one scoop being moved back and forth, a number are set around the periphery of a wheel, (Fig. 51). Like the scoop a paddle wheel is only useful for very low lift pumping, such as flooding paddy fields at no more than about 0.5m height above the water source.

The simplest version is the paddle-wheel in which an operator walks directly on the rim, turning it so that it continuously and steadily scoops up water and deposits it over a low bund, (Fig. 51). In its basic form the paddle wheel is not very efficient since a lot of the water lifted flows back around its edges. Therefore an improved version involves encasing the wheel in a closely fitting box which not only reduces the back-leakage of water but also slightly increases the head through which the device can operate.

Paddle wheels have been mechanized in the past, although they are unusual as water lifting devices today. Many of the windmills used in the Netherlands to dewater large parts of the country drove large paddle wheels, which when mechanized and refined, are usually known as flash wheels. Flash-wheels function best with raked back blades, and the best had measured efficiencies in the range 40-70%. Small straight-bladed paddle-wheels are probably only 10 or 20% efficient, but have the virtue of being simple to build and install in situations were a lot of water needs to be lifted through a small head. They are occasionally used on traditional windpumps, as shown in Fig. 110.

3.6.6 Water Ladders and Dragon-Spine Pumps

The main disadvantage of the paddle wheel just described is that to lift water through a greater height a bigger wheel is needed. The water ladder was developed to get around this problem by taking the paddles and linking them together in an endless belt which can be pulled along an inclined open wooden trough or flume (Fig. 52). The endless belt is driven by a powered sprocket at the discharge end, and passes around a free-wheeling sprocket at the lower end. The lower end of the trough or flume is submerged, so that the moving paddles in the belt, which almost fill the cross section of the flume, push water up it. In many ways this method of water lifting is analagous to a screw pump which also pushes water trapped between the blades of a mechanism up a flume. As with the screw pump there is some back-leakage, but with a well-built unit, this is but a small fraction of the high flow that is established.

The water ladder is still very widely used on small farms in S E Asia for flood irrigation of small fields and paddies from open streams and canals or for pumping sea water into evaporation pans to produce sea salt. In China it is known as a "dragon spine" or dragon wheel" and in Thailand as "rahad". In most cases it is made mainly of wood, and can consequently easily be repaired on-farm. It is one of the most successful traditional, high-flow, low-lift water pumping devices and is particularly applicable to rice production, where large volumes of water are sometimes needed.

On traditional Chinese water ladders, the upper sprocket is normally driven by a long horizontal shaft which traditionally is pedalled by from two to eight people working simultaneously; (Fig. 52). The treadles are spaced on the drive shaft so that one or more of the operators applies full foot pressure at any moment, which helps to smooth the torque output and keep the chain of boards tensioned and running smoothly.

Table 6 SPECIFICATIONS OF CHINESE "DRAGON-SPINE" WATER LIFTS

Name of product	Specifications of products		Weight of water lift (kg)	Volume of timber used (m³)	Factory price (yuan)	Notes
	Length of trough (m)	Dimensions of intake (height x width) (m)				
Single man hand-turning water lift	1.5	0.18 x 0.14	18	0.2	46	
	1.8	0.18 x 0.14	20	0.2	50	
	2.0	0.18 x 0.14	22	0.2	57	
	2.3	0.18 x 0.14	24	0.3	64	
	3.0	0.18 x 0.15	30	0.3	76	
	3.5	0.18 x 0.15	35	0.3	80	
Two men treadling water lift	2.3	0.25 x 0.20	50	0.3	93	
	3.0	0.25 x 0.20	55	0.4	106	
Four men treadling water lift	3.5	0.25 x 0.19	70	0.5	126	
	4.1	0.25 x 0.19	75	0.5	138	
	4.7	0.25 x 0.19	85	0.6	151	
	5.3	0.25 x 0.19	105	0.7	165	
Wind powered water lift of diagonal web member	3.5	0.25 x 0.19	335	1.1	609	Wind sail wheel 4-6m in diameter, coupled with a water lift
	4.1	0.25 x 0.19	345	1.2	622	
	4.7	0.25 x 0.19	350	1.2	635	

Remarks: The products in the table are made by Chengqiao Water Lift and Agricultural Tool Plant, Hangjiang Commune, Putian County.

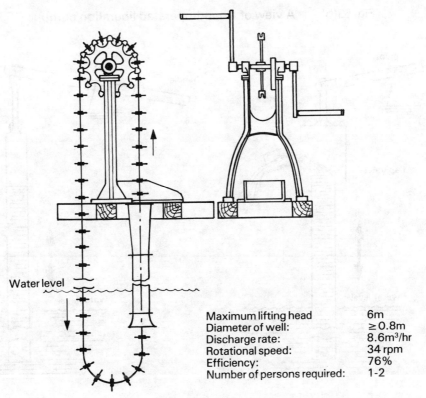

Maximum lifting head: 6m
Diameter of well: ≥ 0.8m
Discharge rate: 8.6m³/hr
Rotational speed: 34 rpm
Efficiency: 76%
Number of persons required: 1-2

Fig. 53(a) Chinese Liberation Wheel chain and washer pump - an animal-powered version is shown in Fig. 96

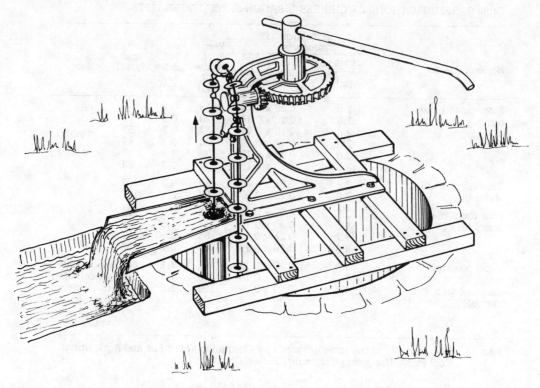

Fig. 53(b)   A view of a hand-operated liberation pump

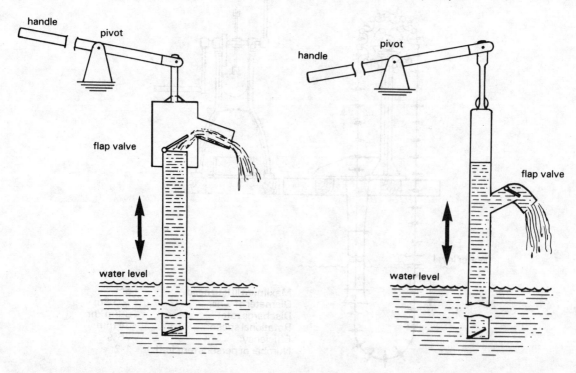

Fig. 54   Flap valve pump   Fig. 55   Joggle pump

Versions of it have been mechanized by using windmills (see Fig. 111), (in Thailand as well as China), or a buffalo sweep (China) or with small petrol (gasoline) engines.

Water ladders range in length from 3 to 8m and in width from 150 to 250mm; lifts seldom exceed 1.0 to 1.2m, but two or more ladders are sometimes used where higher lifts are required. A rough test made in China with a water ladder powered by two teams of four men (one team working and one resting) showed an average capacity of 23m^3/h through a lift of 0.9m [1]. Further details of Chinese water ladders are given in Table 6.

Tests on a traditional wooden water ladder powered by a 2-3hp engine were carried out in Thailand in 1961 [16]. The trough was 190mm deep by 190mm wide and the paddles were 180mm high by 150mm wide and spaced 200mm apart; note that the clearance was quite large, being 20mm each side. The principle findings of this study were:

i. the flow rate is maximized when the submergence of the lower end of the flume is 100%

ii. a paddle spacing to paddle depth ratio of approximately 1.0-1.1 minimizes losses and maximizes output

iii. the sprocket speed has to be kept to less than 80 rpm to avoid excessive wear and frequent breakage

iv. the average efficiency of this device was 40%

It is possible that if a smaller clearance had been used between the paddle edges and the trough, a higher efficiency may have resulted; no doubt the optimum spacing is quite critical. If it were too small, friction would become excessive and possibly cause frequent breakage of the links, while if too large, back-leakage becomes excessive and reduces the overall efficiency.

3.6.7 Chain and Washer or Paternoster Pumps

The origins of this type of pump go back over 2,000 years, and they work on a similar principle to the water ladder just described except that instead of pulling a series of linked paddles through an open inclined flume or trough, a series of linked discs or plugs are pulled through a pipe (Figs. 53, 84 and 96). As with the ladder pump, they lend themselves to human, animal or mechanical prime-movers and are most commonly powered by either a team of two to four people or by a traditional windmill.

As discussed in more detail in the next sections covering the use of human and animal power, a major advantage of this kind of pump is that it requires a steady rotary power input which suits the use of a crank drive with a flywheel, which is a mechanically efficient as well as a comfortable way of applying muscle power. It also also readily matches with engines and other mechanical prime-movers.

The main advantage of the chain and washer pump is that it can be used over a wide range of pumping heads; in this respect it is almost as versatile as the commonly used reciprocating bucket pump as it is applicable on heads ranging from 1m to over 100m. For low lifts, loose fitting washers are good enough to lift water efficiently through the pipe, since back-flow will remain a small and acceptable fraction of total flow. At higher lifts, however, tighter fitting plugs rather than washers are necessary to minimize back-leakage; many materials have been tried, but rubber or leather washers supported by smaller diameter metal discs are commonly used. Most chain and washer pumps have a bell mouth at the base of the riser pipe to guide the washers smoothly into the pipe. With higher lift units where a tighter fit is needed, this is only necessary near the lower end of the riser pipe; therefore the riser pipe usually tapers to a larger diameter for the upper sections to minimize friction (see Fig. 53).

The capacity of a chain and washer pump is a function of the diameter of the riser pipe and of the upward speed of the chain. For example, four men are necessary to power a unit with 6m lift and a 100mm riser tube, [1].

Chain and washer pumps have been, and still are in very widespread use, especially in China, where industrially manufactured pumps of this kind are commonly used and are often known as "Liberation Pumps". They represented in development terms in China a major improvement over more traditional and primitive water lifting techniques and an interim step to modernisation using powered centrifugal pumps. Two to three million Liberation Pumps were used in China at the peak of their use in the 1960s [17]. The following performance characteristics relate to typical chain and washer pumps used in China:

motive power	pumping head	discharge rate	efficiency (pump only)
2 men	6m	5-8m³/h	76%
donkey	12m	7m³/h	68%
3kW(e) motor	15m	40m³/h	65%

This indicates that the chain and washer pump is not only versatile, but also rather more efficient than most pumps. It also has an important characteristic for a positive displacement pump of generally needing less torque to start it than to run it, which makes it relatively easy to match to prime-movers having limited starting torque.

3.7   RECIPROCATING INERTIA (JOGGLE) PUMPS

This range of pumps depend on accelerating a mass of water and then releasing it; in other words, on "throwing" water. They are sometimes known as "inertia" pumps.

As with the other families of pumps so far reviewed, there are both reciprocating inertia pumps, described below, (which are only rarely used) and much more common rotary types which include the centrifugal pump, described in Section 3.8.

3.7.1 Flap Valve Pump

This is an extremely simple type of pump which can readily be improvized; (see Fig. 55). Versions have been made from materials such as bamboo and the dimensions are not critical, so that little precision is needed in building it.

The entire pump and riser pipe are joggled up and down by a hand lever, so that on the up-stroke the flap valve is sucked closed and a column of water is drawn up the pipe, so that when the direction of motion is suddenly reversed the column of water travels with sufficient momentum to push open the flap valve and discharge from the outlet. Clearly a pump of this kind depends on atmospheric pressure to raise the water, so it is limited to pumping lifts of no more than 5-6m.

3.7.2 Resonant Joggle Pump

Fig. 54 shows an improved version of the flap-valve pump. Here there is an air space at the top of the pump which interacts with the column of water by acting as a spring, to absorb energy and then use it to expel water for a greater part of the stroke than is possible with a simple flap-valve pump. This uses exactly the same principle as for an air chamber (see Section 3.5.4).

The joggle pump depends on being worked at the correct speed to make it resonate. An example of a resonant device is a weight hanging from a spring, which will bounce up and down with a natural frequency determined by the stiffness of the spring and the magnitude of the weight. The heavier the weight in relation to the spring stiffness, the slower the natural frequency and vice-versa. If the spring is tweaked regularly, with a frequency close to its natural frequency, then a small regular pull applied once per bounce can produce a large movement quite easily, which is an example of resonance. In exactly the same way, each stroke of a resonant joggle pump makes a column of water of a certain mass bounce on the cushion of air at the top of the column. Depending on the size of the air chamber and the mass of the water, this combination will tend to bounce at a certain resonant frequency. Once it has been started, a pump of this kind needs just a regular "tweak" of the handle at the right frequency to keep the water bouncing. This effect not only improves the overall efficiency but makes it relatively effortless to use. Dunn [20] reports performance figures of 60 to 100 litres/minute lifted through 1.5 to 6m at a frequency of 80 strokes per minute.

It is worth noting that the performance of some reciprocating piston pumps fitted with airchambers (as in Fig. 36 C) can be similarly enhanced if the speed of the pump is adjusted to match the resonant frequency of the water in the pipeline and the "stiffness" of the trapped air in the air-chamber. This is usually only feasible with short pipelines at fairly low heads, as otherwise the natural frequency in most practical cases is far too low to match any reasonable pump speed. If resonance is achieved in such situations the pump will often achieve volumetric efficiencies in the region of 150 to 200%; i.e. approaching twice the swept volume of the pump can be delivered.

This is because the water continues to travel by inertial effects even when the pump piston is moving against the direction of flow, (the valves of course must remain open). As a result, water gets delivered for part of the down stroke as well as on the up stroke. Well-engineered reciprocating systems taking advantage of resonance can achieve high speeds and high efficiencies. Conversely, care may be needed in some situations (such as pumps where there is a reversal of the direction of flow), to avoid resonance effects, as although they can improve the output, they can also impose excessive loads on the pump or on its drive mechanism.

3.8 ROTODYNAMIC PUMPS

3.8.1 Rotodynamic Pumps: Basic Principles

The whole family of so-called rotodynamic pumps depends on propelling water using a spinning impeller or rotor. Two possible mechanisms are used either alone or in combination, so that water is continuously expelled from the impeller by being:

i.   deflected by the impeller blades (in propeller type pumps);

ii. whirled into a circular path so centrifugal force then carries the water away, in the same way a weight on a string when whirled around and released will fly away.

The earliest practical rotodynamic pumps were developed in the 18th and early 19th century, (Fig. 56). Type A in the figure simply throws water outwards and upwards. Type B is actually a suction centrifugal pump and needs priming in order to initiate pumping; a foot valve is provided to prevent the loss of the priming water when the pump stops. A circular casing is provided to collect the output from the impeller at the delivery level. A pump of this kind is extremely inefficient as the water leaves the impeller with a high velocity which is simply dissipated as lost energy. Pump C, the Massachusetts Pump of 1818, had the collector built around a horizontal shaft so that the velocity of the water could be directed up the discharge pipe and carry it to some height; in some respects this is the fore-runner of the modern centrifugal pump which today is the most commonly used mechanically driven type of pump.

3.8.2 Volute, Turbine and Regenerative Centrifugal Pumps

The early pumps just described differed from modern pumps in one important respect; the water left the pump impeller at high speed and was only effectively slowed down by friction, which gives them poor efficiency and poor performance. The application of an important principle, shown in Fig. 57, led to the evolution of efficient rotodynamic pumps; namely that with flowing fluids, velocity can be converted into pressure and vice-versa. The mechanism is to change the cross section of the passage through which water (or any liquid) is flowing. Because water is virtually incompressible, if a given flow is forced to travel through a smaller cross section of passage, it can only do so by flowing faster. However pressure is needed to create the force needed

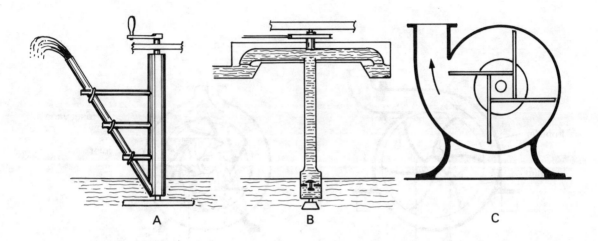

Fig. 56   Early types of centrifugal pumps

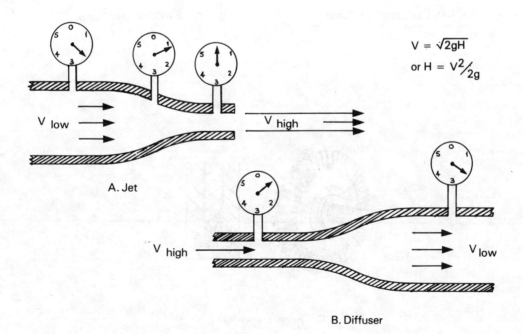

Fig. 57   The relationship between pressure and velocity through both a jet and a diffuser

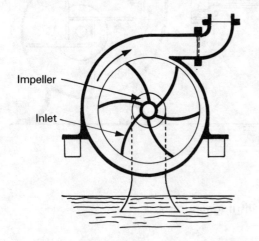

A.—Volute Centrifugal Pump.

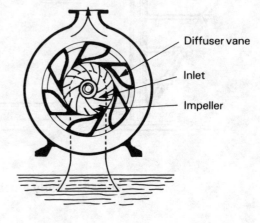

B.—Turbine Centrifugal Pump.

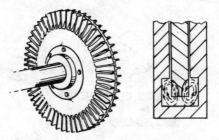

C.—Regenerative Pump

Fig. 58   Centrifugal pump types

to accelerate the mass of water. Conversely, if a flow expands into a larger cross section, it slows down to avoid creating a vacuum and the deceleration of the fluid imposes a force and hence an increase in pressure on the slower moving fluid. It can be shown (if frictional effects are ignored) that if water flows through a duct of varying cross sectional area, then the head of water (or pressure difference) to cause the change in velocity from v_{in} to v_{out}, will be H, where:

$$v^2_{out} - v^2_{in} = \sqrt{2gH}$$

where g is the acceleration due to gravity.

The diagram in Fig. 57 shows how the pressure decreases in a jet as the velocity increases while the reverse occurs in a diffuser which slows water down and increases the pressure. Qualitatively this effect is obvious to most people. From experience, it is well known that pressure is needed to produce a jet of water; the opposite effect, that smoothly slowing down a jet increases the pressure is less obvious.

When this was understood, it became evident that the way to improve a centrifugal pump is to throw the water out of an impeller at high speed (in order to add the maximum energy to the water) and then to pass the water smoothly into a much larger cross section by way of a diffuser in which the cross section changes slowly. In this way, some of the velocity is converted into pressure. A smooth and gradual change of cross-section is essential, any sudden change would create a great deal of turbulence which would dissipate the energy of the water instead of increasing the pressure. There are two main methods of doing this, illustrated in Fig. 58 by diagrams A and B, and a more unusual method shown in C.

Diagram A shows the most common, which is the "volute centrifugal" pump, generally known more simply just as a "centrifugal" pump. Here a spiral casing with an outer snail-shell-shaped channel of gradually increasing cross section draws the output from the impeller tangentially, and smoothly slows it down. This allows the water to leave tangentially through the discharge pipe at reduced velocity, and increased pressure.

Diagram B shows the other main alternative, which is the so-called "turbine centrifugal" or "turbine pump", where a set of smoothly expanding diffuser channels, (six in the example illustrated) serve to slow the water down and raise its pressure in the same way. In the type of turbine pump illustrated, the diffuser channels also deflect the water into a less tangential and more radial path to allow it to flow smoothly into the annular constant cross-section channel surrounding the diffuser ring, from where it discharges at the top.

Diagram C shows the third, lesser known type of centrifugal pump which is usually called a "regenerative pump", but is also sometimes called a "side-chamber pump" or even (wrongly) a "turbine pump". Here an impeller with many radial blades turns in a rectangular sectioned annulus; the blades accelerate the water by creating two strong rotating vortexes which partially interact with the impeller around the rim of the pump for about three-quarters of a revolution; energy is steadily added to the two vortexes each time water passes through the impeller; for those familiar with motor vehicle automatic transmissions the principle is similar to that of the fluid flywheel. When the water leaves the annulus it passes through a diffuser which converts its velocity back into pressure. Regenerative pumps are mentioned mainly for completeness; because they have very close internal clearances they are vulnerable to any suspended grit or dirt and are therefore only normally used with clean water (or other fluids) in situations where their unique characteristics are advantageous. They are generally inappropriate for irrigation duties. Their main advantage is a better capability of delivering water to a higher head than other types of single-stage centrifugal pump.

3.8.3 Rotodynamic Pump Characteristics and Impeller Types

It is not intended to deal with this complex topic in depth, but it is worth running through some of the main aspects relating to pump design to appreciate why pumps are generally quite sensitive to their operating conditions.

All rotodynamic pumps have a characteristic of the kind illustrated in Fig. 16, which gives them a limited range of speeds, flows and heads in which good efficiency can be achieved. Although most pumps will operate over a wider range, if you move far enough from their peak efficiency with any of these parameters, then both the efficiency and output will eventually fall to zero. For example, Fig. 16 shows that if you drive the pump in question with a motor having a maximum speed of say 2000 rpm, there is a maximum flow which can be achieved even at zero head, and similarly there is a head beyond which no flow will occur. The design point is usually at the centre of the area of maximum efficiency.

Since any single rotodynamic pump is quite limited in its operating conditions, manufacturers produce a range of pumps, usually incorporating many common components, to cover a wider range of heads and flows. Because of the limited range of heads and flows any given impeller can handle, a range of sub-sets of different types of impellers has evolved, and it will be shown later there are then variations within each sub-set which can fine tune a pump for different duty requirements. The main sub-sets are shown in Fig. 59, which shows a half-section through the impellers concerned to give an idea of their appearance.

It can be seen that pump impellers impose radial, or axial flow on the water, or some combination of both. Where high flows at low heads are required (which is common with irrigation pumps), the most efficient impeller is an axial flow one (this is similar to a propeller in a pipe) - see Fig. 60. Like a propeller, this depends on lift generated by a moving streamlined blade; since in this case the propeller is fixed in a casing, the reaction moves the water. Conversely, for high heads and low flows a centrifugal (radial flow) impeller is needed with a large ratio between its inlet diameter and its outlet diameter, which produces a large radial flow component, as in the left-most type in Fig. 59. In between these two extremes are mixed flow pumps (see also Figs. 61 and 62) and centrifugal pumps with smaller ratios of discharge to inlet diameter for their impellers. The mixed flow pump has internal blades in the impeller which partially propel the water, as with an axial flow impeller, but the discharge from the impeller is at a greater diameter than the inlet so that some radial flow is involved which adds velocity to the water from centrifugal forces that are generated.

Fig. 59 also shows the efficiency versus the "Specific Speed" of the various impeller sub-sets. Specific Speed is a dimensionless ratio which is

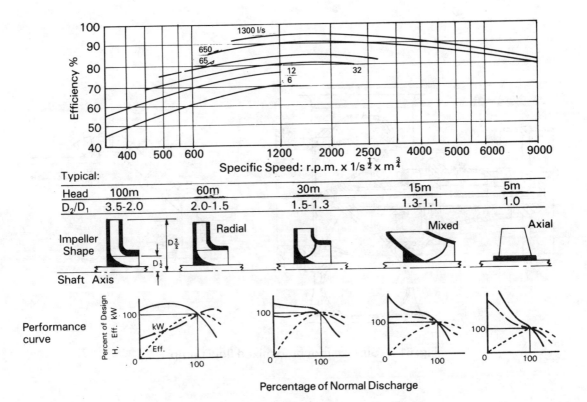

Fig. 59 Typical rotodynamic pump characteristics

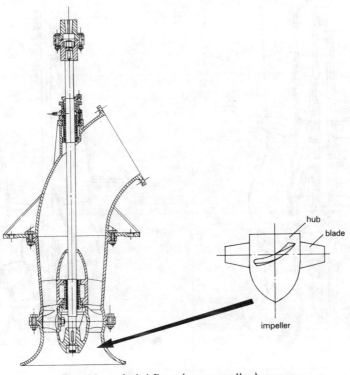

Fig. 60 Axial flow (or propeller) pump

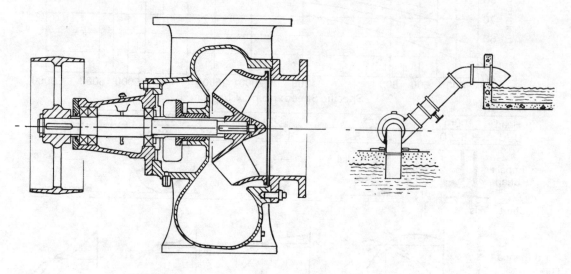

Fig. 61 Surface mounted mixed flow pump

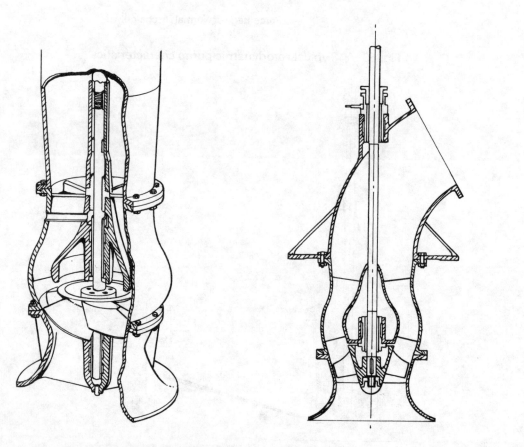

Fig. 62 Submerged mixed flow pump

useful for characterising pump impellers (as well as hydro-turbine rotors or runners). Text books on pump/turbine hydrodynamics cover this topic in greater depth. The Specific Speed is defined as the speed in revolutions per minute at which an impeller would run if reduced in size to deliver 1 litre/sec to a head of 1m and provides a means for comparing and selecting pump impellers and it can be calculated as follows:

$$\text{Specific Speed} \quad N = n \times \frac{\sqrt{Q}}{H^{\frac{3}{4}}}$$

where n is speed in rpm, Q is the pump discharge in litre/sec and H is the head in metres.

Fig. 59 indicates the Specific Speeds which best suit the different impeller sub-sets; e.g. an axial flow impeller is best at flow rates of 500-1,000 litre/sec and has a Specific Speed of 5,000-10,000, at heads of about 5m. Specific Speed can be converted back to actual rpm (n) at any given head (H) and flow (Q) as follows:

$$n = \frac{NH^{\frac{3}{4}}}{Q}$$

where n is in rpm, N is the Specific Speed from Fig. 59, H is the head in metres, and Q is the flow in litres/sec.

The choice of impeller is not only a function of head and flow but of pump size too; smaller low powered pumps of any of these configurations tend to be somewhat less efficient and they also operate best at lower heads than geometrically similar larger versions.

Fig. 59 also indicates the effect on power requirements and efficiency (marked "kW" and "EFF" respectively) of varying the key parameter of head "H", away from the design point. In the case of a centrifugal pump the small diagram shows that increasing the head reduces the power demand, while in the case of an axial-flow pump, increasing the head increases the power demand. Paradoxically, reducing the head from the design head on a centrifugal pump increases the power demand; the reason for this is that decreasing the head by, say, 10% can increase the flow by 25% - the efficiency may also go down by 10%, and since the power requirement is head times flow divided by efficiency, the new power demand will change from:

$$\frac{H \times Q}{EFF} \quad \text{to} \quad \frac{0.9H \times 0.25Q}{0.9EFF}$$

the ratio of these is 1:1.25, so the power demand will be increased by 25% in this case. Therefore, varying the conditions under which a pump operates away from the design point can have an unexpected and sometimes drastic effect. The use of pumps off their design point is a common cause of gross inefficiency and wasted fuel.

3.8.4 Axial-Flow (Propeller) Pumps

As already explained, an axial-flow (or "propeller") pump propels water by the reaction to lift forces produced by rotating its blades. This action both pushes the water past the rotor or impeller and also imparts a spin to the water which if left uncorrected would represent wasted energy, since it will increase the friction and turbulence without helping the flow of water down the pipe. Axial flow pumps therefore usually have fixed guide vanes, which are angled so as to straighten the flow and convert the spin component of velocity into extra pressure, in much the same way as with a diffuser in a centrifugal pump. Fig. 60 shows a typical axial flow pump of this kind, in which the guide vanes, just above the impeller, also serve a second structural purpose of housing a large plain bearing, which positions the shaft centrally. This bearing is usually water lubricated and has features in common with the stern gear of an inboard-engined motor boat.

Axial flow pumps are generally manufactured to handle flows in the range 150 to 1,500m³/h for vertically mounted applications, usually with heads in the range 1.5-3.0m. By adding additional stages (i.e. two or more impellers on the same shaft) extra lift up to 10m or so can be engineered.

Because pumps of this kind are designed for very large flows at low heads, it is normal to form the "pipes" in concrete as illustrated, to avoid the high cost of large diameter steel pipes. Most axial flow pumps are large scale devices, which involve significant civil works in their installation, and which would generally only be applicable on the largest land-holdings addressed by this publication. They are generally mainly used in canal irrigation schemes where large volumes of water must be lifted 2-3m, typically from a main canal to a feeder canal.

Small scale propeller pumps are quite successfully improvised but not usually manufactured; ordinary boat propellers mounted on a long shaft have been used for flooding rice paddies in parts of southeast Asia. The International Rice Research Institute (IRRI) has developed this concept into a properly engineered, portable high volume pumping system, (see Fig. 63); it is designed to be manufactured in small machine shops and is claimed to deliver up to 180m³/h at heads in the range 1-4m. This pump requires a 5hp (3kW) engine or electric motor capable of driving its shaft at 3,000rpm; its length is 3.7m, the discharge tube is 150mm in diameter and the overall mass without the prime mover fitted is 45kg.

3.8.5 Mixed-Flow Pumps

The mixed-flow pump, as its name suggests, involves something of both axial and centrifugal pumps and in the irrigation context can often represent a useful compromise to avoid the limited lift of an axial flow pump, but still achieve higher efficiency and larger flow rates than a centrifugal volute pump. Also, axial flow pumps generally cannot sustain any suction lift, but mixed-flow pumps can, although of course they are not self-priming.

Fig. 61 shows a surface mounted, suction mixed-flow pump and its installation. Here the swirl imparted by the rotation of the impeller is recovered by delivering the water into a snail-shell volute or diffuser, identical in principle to that of a centrifugal volute pump.

An alternative arrangement more akin to an axial flow pump is shown in Fig. 62. Here what is often called a "bowl" casing is used, so that the flow spreads radially through the impeller, and then converges axially through fixed guide vanes which remove the swirl and thereby, exactly as with axial flow pumps, add to the efficiency. Pumps of this kind are installed submerged,

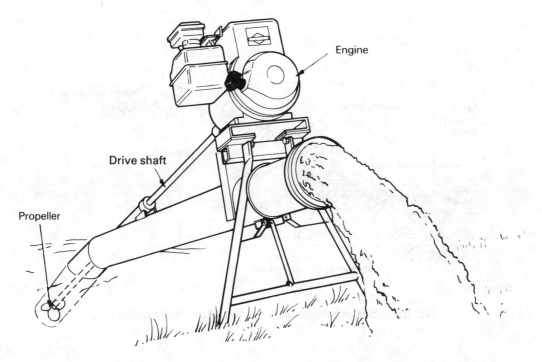

Fig. 63   Portable axial flow pump (IRRI)

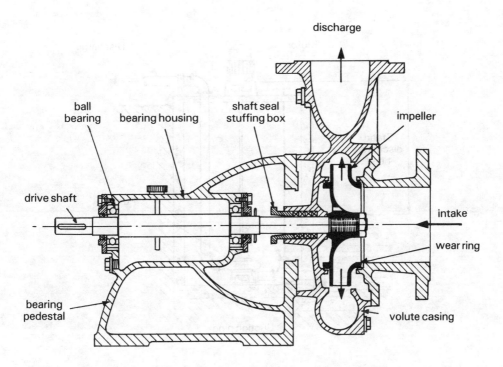

Fig. 64   Typical surface mounted pedestal centrifugal pump

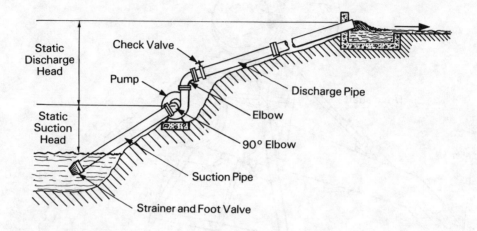

Fig. 65 Surface mounted centrifugal pump installation

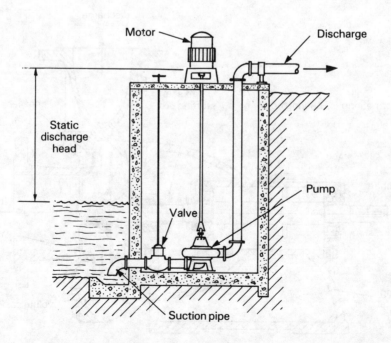

Fig. 66 Below-surface (sump) centrifugal pump installation

which avoids the priming problems that can afflict large surface suction rotodynamic pumps such as in Fig. 61. The "bowl" mixed-flow pump is sometime called a "turbine" pump, and it is in fact analagous to the centrifugal turbine pump described earlier; the passage through the rotor reduces in cross-section and serves to accelerate the water and impart energy to it, while the fixed guide vanes are designed as a diffuser to convert speed into pressure and thereby increase both the pumping head and the efficiency. A number of bowl pumps can be stacked on the same shaft to make a multi-stage turbine pump, and these are quite commonly used as borehole pumps due to their long narrow configuration. Mixed-flow bowl pumps typically operate with flows from 200-12,000m³/h over heads from 2-10m. Multiple stage versions are often used at heads of up to about 40m.

3.8.6 Centrifugal Pumps

i. **Horizontal shaft centrifugal pump construction**

These are by far the most common generic type of electric or engine powered pump for small to medium sized irrigation applications. Fig. 64 shows a typical mass-produced volute-centrifugal pump in cross section. In this type of pump the casing and frame are usually cast iron or cast steel, while the impeller may be bronze or steel. Critical parts of the pump are the edges of the entry and exit to the impeller as a major source of loss is back-leakage from the exit of the impeller around the front of it to the entry. To prevent this, good quality pumps, including the one in the diagram, have a closely fitting wear ring fitted into the casing around the front rim of the impeller; this is subject to some wear by grit or particulate matter in the water and can be replaced when the clearance becomes large enough to cause significant loss of performance. However, many farmers probably do not recognize wear of this component as being serious and simply compensate by either driving the pump faster or for longer each day, both of which waste fuel or electricity. Another wearing part is a stuffing box packing where the drive shaft emerges from the back of the impeller casing. This needs to be periodically tightened to minimize leakage, although excessive tightening increases wear of the packing. The packing is usually graphited asbestos, although graphited PTFE is more effective if available. The back of the pump consists of a bearing pedestal and housing enclosing two deep-groove ball-bearings. This particular pump is oil lubricated, it has a filler, dip-stick and drain plug. Routine maintenance involves occasional changes of oil, plus more frequent checks on the oil level. Failure to do this leads to bearing failure, which if neglected for any time allows the shaft to whirl and damage the impeller edges.

ii. **Centrifugal pump installations**

Figs. 65 and 66 show two alternative typical low lift centrifugal pump installations; the simplest is the suction installation of Fig. 65. As mentioned earlier in Section 2.1.5, centrifugal pumps are limited to a maximum in practice of about 4-5m suction lift at sea level (reducing to around 2m suction lift at an altitude of 2,000m, and further reduced if a significant length of suction pipe is involved; otherwise problems are almost certain to be experienced in priming the pump, retaining its prime, etc. A foot valve is a vital part of any such installation as otherwise the moment the pump stops or slows down, all the water in the pipeline will run back through the pump making it impossible to restart the pump unless the pipeline is first refilled. Also, if water flows back through the pump, it can run backwards and possibly damage the electrical system.

If the delivery pipeline is long, it is also important to have another check valve (non-return valve) at the pump discharge to the pipeline. The reason for this is that if for any reason the pump suddenly stops, the flow will continue until the pressure drops enough to cause cavitation in the line; when the upward momentum of the water is exhausted, the flow reverses and the cavitation bubbles implode creating severe water hammer. Further severe water hammer occurs when the flow reverses causing the footvalve to slam shut. The impact of such events has been known to burst a centrifugal pump's casing. The discharge check valve therefore protects the pump from any such back surge down the pipeline.

In many cases there is no surface mounting position low enough to permit suction pumping. In such cases centrifugal pumps are often placed in a sump or pit where the suction head will be small, or even as in Fig. 66 where the pump is located below the water level. In the situation illustrated a long shaft is used to drive the pump from a surface mounted electric motor; (to keep the motor and electrical equipment above any possible flood level).

iii. Centrifugal pump impeller variations

The component that more than anything else dictates a centrifugal pump's characteristics is its impeller. Fig. 67 shows some typical forms of impeller construction. Although the shape of an impeller is important, the ratio of impeller exit area to impeller eye area is also critical (i.e. the change of cross section for the flow through the impeller), and so is the ratio of the exit diameter to the inlet diameter. A and B in the figure are both open impellers, while C and D are shrouded impellers. Open impellers are less efficient than shrouded ones, (because there is more scope for back leakage and there is also more friction and turbulence caused by the motion of the open blades close to the fixed casing), but open impellers are less prone to clogging by mud or weeds. But shrouded impellers are considerably more robust and less inclined to be damaged by stones or other foreign bodies passing through. Arguably, open impellers are less expensive to manufacture, so they tend to be used on cheaper and less efficient pumps; shrouded impellers are generally superior where efficiency and good performance are important.

Also in Fig. 67, A and C are impellers for a single-suction pump, while B and D are for a double-suction pump in which water is drawn in symmetrically from both sides of the impeller. The main advantage of a double-suction arrangement is that there is little or no end thrust on the pump shaft, but double suction pumps are more complicated and expensive and are uncommon in small and medium pump sizes.

The shape of the impeller blades is also of importance. Some factors tend to flatten the HQ curve for a given speed of rotation, while others steepen it. Fig. 68 shows the effect of backward raked, radial and forward raked blade tips; the flattest curve is obtained with the first type, while the last type actually produces a maximum head at the design point. Generally the flatter the HQ curve, the higher the efficiency, but the faster the impeller has to be driven to achieve a given head. Therefore impellers

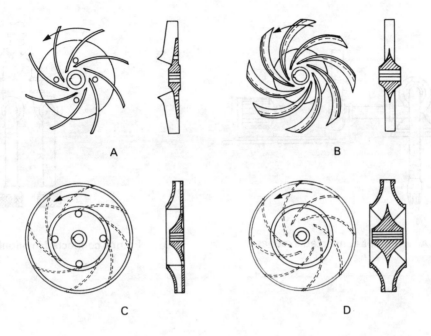

Fig. 67 Various types of centrifugal pump impellers

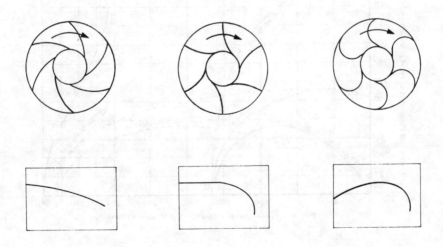

Fig. 68 Effect of direction of curvature of vanes of centrifugal pump impellers

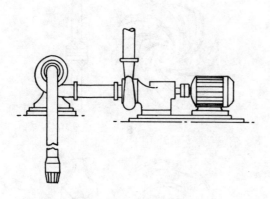

A   Pumps connected in series

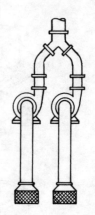

B   Pumps connected in parallel

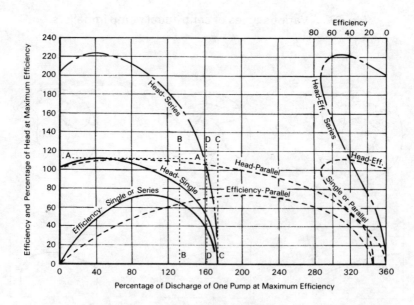

C   Characteristics of two similar centrifugal pumps connected in parallel and in series

Fig. 69   Combining centrifugal pumps in series or parallel

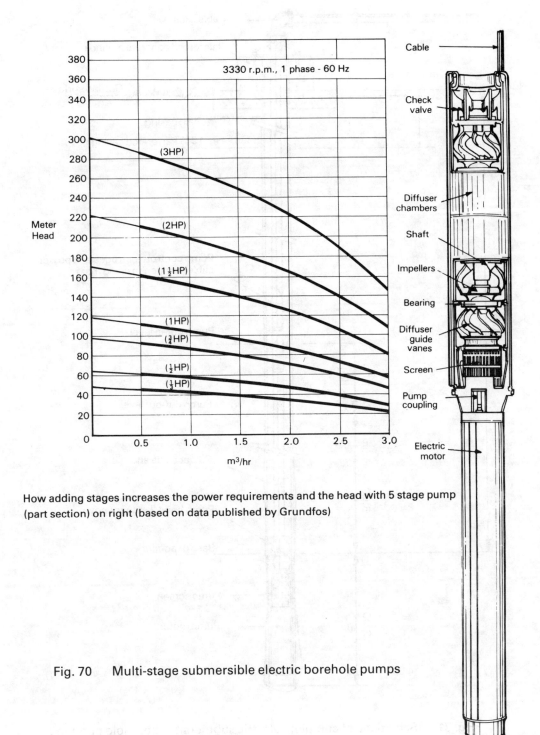

How adding stages increases the power requirements and the head with 5 stage pump (part section) on right (based on data published by Grundfos)

Fig. 70 Multi-stage submersible electric borehole pumps

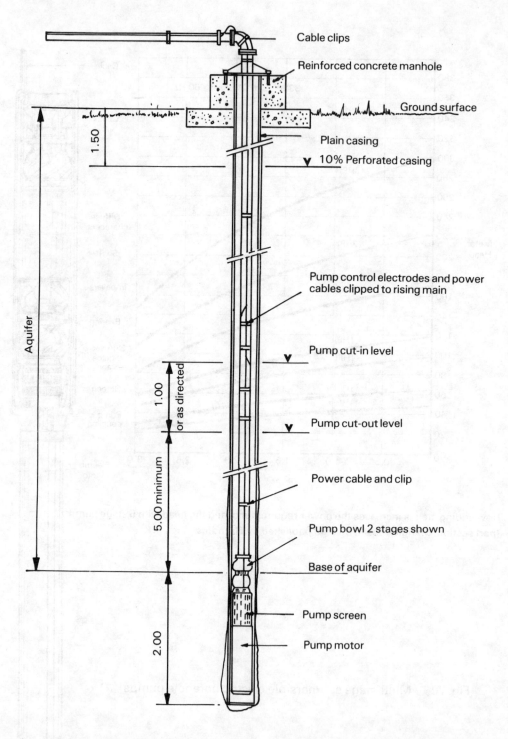

Fig. 71   Schematic of complete electric submersible borehole pumping installation

producing the most humped characteristics tend to be used when a high head is needed for a given speed, but at some cost in reduced efficiency.

iv. Series and parallel operation of centrifugal pumps

Where a higher head is needed than can be achieved with a single pump, two can be connected in series as in Fig. 69 A, and similarly, if a greater output is needed, two centrifugal pumps may be connected in parallel as in Fig. 69 B. The effects of these arrangements on the pump characteristics are illustrated in Fig. 69 C, which shows the changes in head, discharge and efficiency that occur as a percentage of those for a single pump operating at its design point. It is clear that series connection of pumps has no effect on efficiency or discharge but doubles the effective head. Parallel operation does not however normally double the discharge compared to a single pump, because the extra flow usually causes a slight increase in total head (due to pipe friction), which will move the operating point enough to prevent obtaining double the flow of a single pump.

3.8.7 Multi-stage and Borehole Rotodynamic Pumps

Where high heads are needed, the primary means to achieve this with a single impeller centrifugal pump are either to drive the impeller faster or to increase its diameter. In the end there are practical limits to what can be done in this way, so that either single impeller pumps can be connected in series, or a more practical solution is to use a multiple impeller pump in which the output from one impeller feeds directly, through suitable passages in the casing, to the next, mounted on the same shaft. Fig. 70 shows a 5 stage borehole pump (where limitations on the impeller diameter are caused by the borehole, making multi-staging an essential means to obtain adequate heads). Fig. 144 includes a three stage centrifugal pump, coupled to a turbine as a prime-mover, as another example of multi-staging.

Surface-mounted multi-stage pumps are probably only likely to be of relevance to irrigation in mountainous areas since there are few situations elsewhere where surface water needs to be pumped through a high head. More important from the irrigation point of view is the vertical shaft multi-stage submersible borehole pump which has an integral submerged electric motor directly coupled to the pump below the pump as in the example of Fig. 70. However, it is possible to get bare-shaft multi-stage borehole pumps in which the pump is driven from the surface via a long drive shaft supported by spider bearings at regular intervals down the rising main; see Fig. 134 (b) or with the motor arranged as for the centrifugal pump in Fig. 66, but with a vertically mounted multi-stage pump in either a sump or well.

In recent years numerous, reliable, submersible electric pumps have evolved; Fig. 70. Section 4.6 discusses in more detail the electrical implications and design features of this kind of motor. Extra pump stages can be fitted quite easily to produce a range of pumps to cover a wide spectrum of operating conditions. The pump in Fig. 70 is a 5 stage mixed-flow type, and the same figure also shows how, simply by adding extra stages (with increasingly powerful motors) a whole family of pumps can be created capable in the example illustrated of lifting water from around 40m with the smallest unit to around 245m with the most powerful; the efficiency and flow will be similar for all of these options. Only the head and the power rating will vary in proportion to the number of stages fitted.

Finally, Fig. 71 and Fig. 134 (a) show borehole installations with submersible electric pumps. The pump in Fig. 71 has level sensing electrodes clipped to the rising main, which can automatically switch it off if the level falls too low.

3.8.8 Self-priming Rotodynamic Pumps

Rotodynamic pumps, of any kind, will only start to pump if their impellers are flooded with water prior to start-up. Obviously the one certain way to avoid any problem is to submerge the pump in the water source, but this is not always practical or convenient. This applies especially to portable pump sets, which are often important for irrigation, but which obviously need to be drained and re-primed every time they are moved to a new site.

If sufficient water is present in the pump casing, then even if the suction pipe is empty, suction will be created and water can be lifted. A variety of methods are used to fill rotodynamic pumps when they are mounted above the water level. It is, however, most important to note that if the suction line is empty but the delivery line is full, it may be necessary to drain the delivery line in order to remove the back pressure on the pump, to enable it to be primed. Otherwise it will be difficult if not impossible to flush out the air in the system. One way to achieve this is to fit a branch with a hand valve on it at the discharge, which can allow the pump to be "bled" by providing an easy exit for the air in the system.

The most basic method of priming is to rely on the footvalve to keep water in the system. The system has to be filled initially by pouring water into the pipes from a bucket; after that it is hoped that the footvalve will keep water in the system even after the pump is not used for some time. In many cases this is a vain hope, as footvalves quite often leak, especially if mud or grit is present in the water and settles between the valve and its seat when it attempts to close. Apart from the nuisance value when a pump loses its prime, many pumps suffer serious damage if run for any length of time while dry, as the internal seals and rubbing faces depend on water lubrication and will wear out quickly when run dry. Also, a pump running dry will tend to overheat; this will melt the grease in the bearings and cause it to leak out, and can also destroy seals, plastic components or other items with low temperature tolerance.

The two most common methods for priming surface-mounted, engine driven suction centrifugal pumps are either by using a small hand pump on the delivery line as illustrated in Fig. 72, (this shows a diaphragm priming pump which has particularly good suction capabilities) or an "exhaust ejector" may be used; here suction is developed by a high velocity jet of exhaust from the engine, using similar principles to those illustrated in Fig. 57 and described in more detail in Section 3.8.9 which follows.

Several alternatie methods of priming surface suction pumps may be commonly improvised. For example, a large container of water may be mounted above the pump level so water can be transferred between the pump and the tank via a branch from the delivery line with a valve in it. Then when the pump has to be restarted after the pipe-line has drained, the valve can be opened

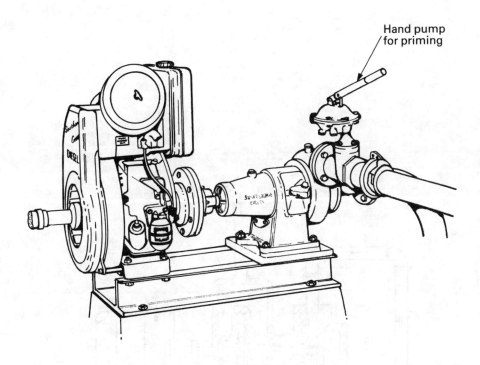

Fig. 72 Direct-coupled air-cooled diesel engine and pump installation with hand-operated diaphragm pump for priming

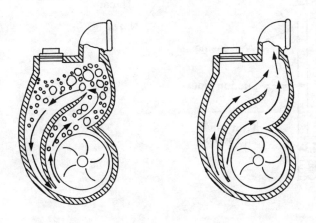

Fig. 73 Self-priming centrifugal pump

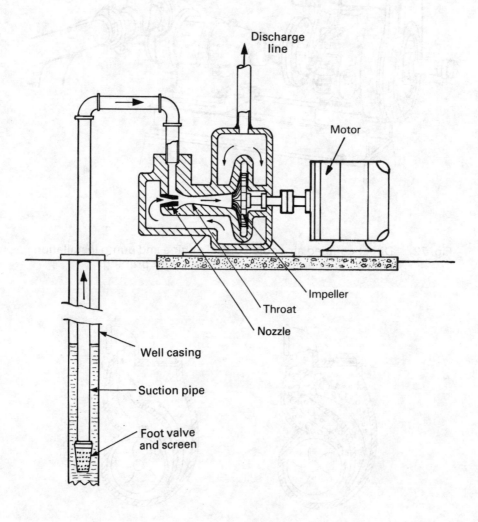

Fig. 74 Schematic of a surface-suction jet pump

to drain the tank into the pump and suction line. Even the worst footvalves leak slowly enough to enable the system to be started, after which the tank can be refilled by the pump so as to be ready for the next start.

Alternatively, a large container can be included in the suction line, mounted above the level of the pump, which will always trap enough water in it to allow the pump to pull enough of a vacuum to refill the complete suction line. Care is needed in designing an installation of this kind, to avoid introducing air-locks in the suction line.

Yet another simple method to use, but only if the delivery line is long enough to carry a sufficient supply of water, is to fit a hand-valve immediately after the pump discharge (instead of a non-return valve) so that when the pump is turned off, the valve can be manually closed. Then the opening of this valve will refill the pump from the delivery line to ensure it is flooded on restarting.

Sometimes the most reliable arrangement is to use a special "self-priming" centrifugal pump (Fig. 73). Here, the pump has an enlarged upper casing with a baffle in it. When the pump and suction line are empty, the pump casing has to be filled with water from a bucket through the filler plug visible on top. Then when the pump is started, the water in the casing is thrown up towards the discharge and an eye is formed at the hub of the impeller which is at low pressure; until water is drawn up the suction pipe the water discharged from the top of the pump tends to fall back around the baffle and some of the entrained air carries on up the empty discharge pipe. The air which is discharged is replaced by water drawn up the suction pipe, until eventually the suction pipe fills completely and the air bubble in the eye of the impeller is blown out of the discharge pipe. Once all the air has been expelled, water ceases to circulate within the pump and both channels act as discharge channels. A check valve is fitted to the inlet of the pump so that when the pump is stopped it remains full of water. Then even if the foot valve on the suction line leaks and the suction line empties, the water trapped in the casing of the pump will allow the same self-priming function as described earlier to suck water up the suction line. Hence, pumps of this kind only need to be manually filled with water when first starting up after the entire system has been drained.

3.8.9 Self-Priming Jet Pumps

An alternative type of self-priming centrifugal pump uses the fact that if water is speeded up through a jet, it causes a drop in pressure (see Section 3.8.2). Here the pump is fitted into a secondary casing which contains water at discharge pressure, (see Fig. 74). A proportion of the water from this chamber is bled back to a nozzle fitted into the suction end of the pump casing and directed into the eye of the impeller. Once the pump has been used once (having been manually primed initially) it remains full of water so that on start up the pump circulates water from the discharge through the jet and back into the suction side. As before, air is sucked through and bubbles out of the discharge, while (until the pump primes) the water falls back and recirculates. The jet causes low pressure in the suction line and entrains air which goes through the impeller and is discharged, hence water is gradually drawn up the suction line. As soon as all the air is expelled from the system, most of the discharge goes up the discharge line, but a proportion is fed back to the nozzle and increases the suction considerably compared with the effect of a centrifugal impeller on its own. Therefore, this kind of pump not only pulls a higher suction lift than normal, but the pump can reliably run on "snore" (i.e. sucking a mixture of air and water without losing its prime). This makes it useful in situations where shallow water is being suction pumped and it is difficult to obtain sufficient submergence of the footvalve, or where a water source may occasionally be pumped dry.

This jet pump principle can also be applied to boreholes as indicated in Fig. 75. An arrangement like this allows a surface-mounted pump and motor to "suck" water from depths of around 10-20m; the diffuser after the jet serves to raise the pressure in the rising main and prevent cavitation. Although the jet circuit commonly needs 1.5-2 times the flow being delivered, and is consequently a source of significant power loss, pumps like this are sometimes useful for lifting sandy or muddy water as they are not so easily clogged as a submerged pump. In such cases a settling tank is provided on the surface between the pump suction and the jet pump discharge to allow the pump to draw clearer water.

The disadvantages of jet pumps are, first, greater complexity and therefore cost, and second, reduced efficiency since power is used in pumping water through the jet, (although some of this power is recovered by the pumping effect of the jet). Obviously it is better to use a conventional centrifugal pump in a situation with little or no suction lift, but where suction pumping is essential, then a self-priming pump of this kind can offer a successful solution.

3.9 AIR LIFT PUMPS

The primary virtue of air lift pumps is that they are extremely simple. A rising main, which is submerged in a well so that more of it is below the water level than above it, has compressed air blown into it at its lowest point (see Fig. 76). The compressed air produces a froth of air and water, which has a lower density than water and consequently rises to the surface. The compressed air is usually produced by an engine driven air compressor, but windmill powered air compressors are also used. The principle of it is that an air/water froth, having as little as half the density of water, will rise to a height above the water level in the well approximately equal to the immersed depth of the rising main. The greater the ratio of the submergance of the rising main to the static head, the more froth will be discharged for a given supply of air and hence the more efficient an air lift pump will be. Therefore, when used in a borehole, the borehole needs to be drilled to a depth more than twice the depth of the static water level to allow adequate submergence.

The main advantage of the air lift pump is that there are no mechanical below-ground components, so it is essentially simple and reliable and can easily handle sandy or gritty water. The disadvantages are rather severe; first, it is inefficient as a pump, probably no better, at best, than 20-30% in terms of compressed air energy to hydraulic output energy, and this is compounded by the fact that air compressors are also generally inefficient. Therefore the running costs of an air lift pump will be very high in energy terms. Second, it usually requires a borehole to be drilled considerably deeper than otherwise would be necessary in order to obtain enough submergence, and this is generally a costly exercise. This problem is

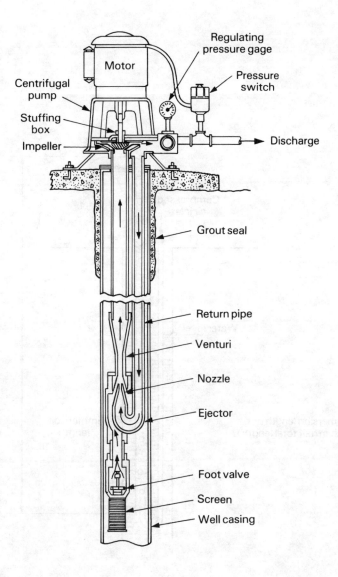

Fig. 75   Borehole jet pump installation

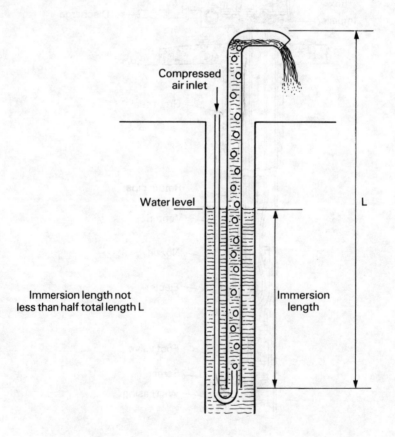

Fig. 76   Air lift pump (schematic)

obviously less serious for low head applications where the extra depth required would be small, or where a borehole needs to be drilled to a considerable depth below the static water level anyway to obtain sufficient inflow of water.

3.10 IMPULSE (WATER HAMMER) DEVICES

These devices apply the energy of falling water to lift a fraction of the flow to a higher level than the source. The principle they work by is to let the water from the source flow down a pipe and then to create sudden pressure rises by intermittently letting a valve in the pipe slam shut. This causes a "water hammer" effect which results in a sudden sharp rise in water pressure sufficient to carry a small proportion of the supply to a considerably higher level.

They therefore are applicable mainly in hilly regions in situations where there is a stream or river flowing quite steeply down a valley floor, and areas that could be irrigated which are above the level that can be commanded by small channels contoured to provide a gravity supply.

The only practical example of a pump using this principle is the hydraulic ram pump, or "hydram", which is in effect a combined water-powered prime mover and pump. The hydraulic ram pump is mechanically extremely simple, robust and ultra reliable. It can also be reasonably efficient. However in most cases the output is rather small (in the region of 1-3 litre/sec) and they are therefore best suited for irrigating small-holdings or single terrace fields, seedlings in nurseries, etc.

Hydraulic rams are described more fully in Section 4.9.3 dealing with water powered pumping devices.

3.11 GRAVITY DEVICES

3.11.1 Syphons

Strictly speaking syphons are not water-lifting devices, since, after flowing through a syphon, water finishes at a lower level than it started. However syphons can lift water over obstructions at a higher level than the source and they are therefore potentially useful in irrigation. They also have a reputation for being troublesome, and their principles are often not well understood, so it is worth giving them a brief review.

Fig. 77 A to C shows various syphon arrangements. Syphons are limited to lifts of about 5m at sea level for exactly the same reasons relating to suction lift for pumps. The main problem with syphons is that due to the low pressure at the uppermost point, air can come out of solution and form a bubble, which initially causes an obstruction and reduces the flow of water, and which can grow sufficiently to form an airlock which stops the flow. Therefore, the syphon pipe, which is entirely at a sub-atmospheric pressure, must be completely air-tight. Also, in general, the faster the flow, the lower the lift and the more perfect the joints, the less trouble there is likely to be with air locks.

Starting syphons off can also present problems. The simplest syphons can be short lengths of flexible plastic hose which may typically be used to irrigate a plot by carrying water from a conveyance channel over a low bund; it is well known that all that needs to be done is to fill the length of hose completely by submerging it in the channel and then one end can be covered by hand usually and lifted over the bund, to allow syphoning to start. Obviously, with bigger syphons, which are often needed when there is an obstruction which cannot easily be bored through or removed, or where there is a risk of leakage from a dam or earth bund if a pipe is buried in it, simple techniques like this cannot be used.

In Fig. 77 A, a non-return valve or foot-valve is provided on the intake side of the syphon, and an ordinary gate valve or other hand-valve at the discharge end. There is a tapping at the highest point of the syphon which can be isolated, again with a small hand valve. If the discharge hand valve is closed and the top valve opened, it is possible to fill the syphon completely with water; the filler valve is then closed, the discharge valve opened and syphoning will commence.

Diagram B is similar to A except that instead of filling the syphon with water to remove the air, a vacuum pump is provided which will draw out the air. Obviously this is done with the discharge valve closed. The vacuum pump can be a hand pump, or it could be a small industrial vacuum pump. Once the air is removed, the discharge valve can be opened to initiate syphoning.

Diagram C shows a so-called "reverse" syphon, used for example where a raised irrigation channel needs to cross a road. Reverse syphons operate at higher than atmospheric pressure and there is no theoretical limit to how deep they can go, other than that the pipes must withstand the hydrostatic pressure and that the outflow must be sufficiently lower than the inflow to produce the necessary hydraulic gradient to ensure gravity flow.

3.11.2 Qanats and Foggara

Qanats, as they are known in Farsi or Foggara (in Arabic), are "man-made springs" which bring water out to the surface above the local water table, but by using gravity. Like syphons they are not strictly water lifting devices, but they do offer an option in lieu of lifting water from a well or borehole in order to provide irrigation. They have been used successfully for 2,000 years or more in Iran, and for many centuries in Afghanistan, much of the Middle East and parts of North Africa.

Fig. 78 shows a cross-section through a qanat; it can be seen that the principle used exploits the fact that the water table commonly rises under higher ground. Therefore, it is possible to excavate a slightly upward sloping tunnel until it intercepts the water table under higher ground possibly at some distance from the area to be irrigated. It is exactly as if you could take a conventional tube well and gradually tip it over until the mouth was below the level of the water table, when, clearly water would flow out of it continuously and without any need for pumping.

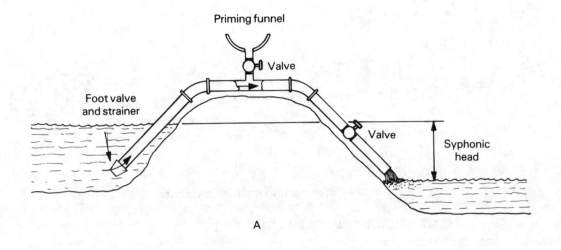
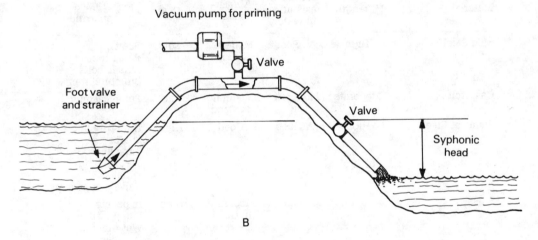
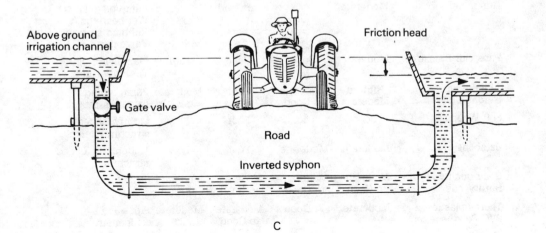

Fig. 77   Syphon arrangements

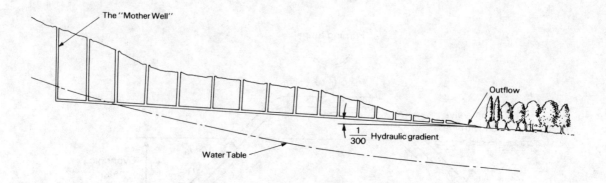

Fig. 78   Cross-section through a Qanat

Table 7   RELATIVE MERITS OF MATERIALS FOR PUMPS

Material	Strength	Corrosion resistance in water	Abrasion resistance	Cost	Typical application
Mild Steel	High	V. Poor	Moderate to Good	Low	Shafts Pump rods nuts & bolts Structural items
Cast Iron	Moderate	Moderate	Moderate to Good	Low	Pump casings
Stainless Steel	High	V. Good	Good	High	Nuts & bolts Shafts Impellers Wet rubbing surfaces Valve components
Brass	Moderate	Good	Moderate to Good	High	Impellers Pump cylinders Wet rubbing surfaces
Bronze/ Gun metal	High to Moderate	V. Good	Moderate to Good	High	Impellers Pump pistons Wet bearings & Rubbing surfaces Valve parts
Phosphor Bronze	Moderate	Good	Good	High	Plain bearings & Thrust washers
Aluminium & light alloys	High to Moderate	Moderate to Good	Poor	Moderate to High	Pump casings Irrigation pipes
Soft Woods	Poor	Poor	Poor	Low	Lightly loaded structural items
Bamboos	Moderate	Moderate	Poor	Low	Moderately loaded structures
Good quality Hardwoods	Moderate to Good	Moderate to Good	Moderate to Good	Moderate to High	Structures
Thermoplastics PVC Polythene, etc.	Moderate	V. Good	Moderate to Good	Moderate	Pipes and components
Thermoplastics, Filled plastics & Composites	High to Moderate	Generally Good	Generally Good	Moderate to High	Pump casings Components Bearings

Qanats are typically from one to as much as 50 kilometres long, (some of the longest are in Iran near Isfahan). They are excavated by sinking wells every 50 to 100m and then digging horizontally to join the bases of the wells, starting from the outflow point. Traditional techniques are used, involving the use of simple hand tools, combined with sophisticated surveying and tunnelling skills. Many decades are sometimes needed to construct a long qanat, but once completed they can supply water at little cost for centuries. The surface appearance of a qanat is distinctive, consisting of a row of low crater-like earth bunds (or sometimes a low brick wall) surrounding each well opening; this is to prevent flash floods from pouring down the well and washing the sides away. The outflow from a qanat usually runs into a cultivated oasis in the desert, resulting from the endless supply of water.

Efforts have been made in Iran to mechanize qanat construction, but without great success, although in some cases qanats are combined with engine powered lift pumps in that the qanat carries water more or less horizontally from under a nearby hill possessing a raised water table to a point on level ground above the local water table but below the surface, where a cistern is formed in the ground. A diesel pump is then positioned on a ledge above the cistern to lift the water to the surface.

3.12 MATERIALS FOR WATER-LIFTING DEVICES

This is a complex technical subject if discussed fully, but it is worth briefly setting out some of the advantages and disadvantages of different materials that are commonly used, as an aid to appraising the specification of different equipment.

Four main considerations apply for construction materials used for pumping water:

a. strength; stressed components need to be able to function over a long period of time without either failing through overload or, more likely, through fatigue;

b. corrosion resistance and general ability to coexist under wet conditions;

c. resistance to wear and abrasion is important for components that rub or slide or which are in contact with flowing water if any particulate matter is likely to be suspended in the water;

d. cost.

As in most branches of engineering, nature has not been kind enough to offer materials which simultaneously satisfy all these requirements completely; invariably compromises are necessary. The important point is to be aware of these and to judge whether they are the right compromises for the application of interest.

It is worth reviewing briefly the pros and cons of various different materials which feature frequently in pumps and water lifts; these are also summarised in Table 7.

i   <u>Ferrous Metals</u>

Most ferrous, or iron based, materials are subject to corrosion problems, but to compensate, they are perhaps the most familiar low-cost "strong material" that is widely available. Generally speaking iron and steel are best suited for use in structural components where strength is important but a surface coating of rust will not cause serious problems.

Ordinary mild steel is one of the most susceptible to corrosion. Iron and steel castings, except where they have been machined, are generally partially protected by black-iron oxide which forms when the casting is still hot. There are several methods to protect steels from corrosion, including conventional paints, various modern corrosion inhibitors which chemically bond with the surface of the metal and inhibit corrosion, various forms of plating and metallic coatings such as zinc (galvanizing) and cadmium plating. Various steel chromium and nickel based alloys, the so-called stainless steels, are also resistant to oxidation and corrosion, but they are not cheap.

Stainless steels do make a useful alternative to brasses and bronzes, but they are very difficult to machine and to work and therefore most pump manufacturers prefer non-ferrous corrosion resistant alloys. One important application for stainless steel is as nuts and bolts in situations where mild steel nuts and bolts readily corrode; stainless steel nuts and bolts are expensive compared with mild steel ones, but cheap in terms of time saved in the field on items that regularly need to be dismantled in wet conditions for maintenance or replacement.

A primary mechanism for corrosion of steel in wet conditions is if the steel is in combination with nobler metals, (e.g. copper), and there is an electrical link between them while both metals are in contact with water. This can encourage what is known as electrolytic corrosion, especially if the water has a significant mineral content which will generally increase its conductivity.

Therefore, ferrous components ought to be well protected from corrosion and generally are best suited as structural items not having any "high quality" surfaces in contact with water. An example of a bad use for iron, where it sometimes is applied, is as cast iron pump cylinders. Here the internal surface will often keep in quite good condition so long as the pump is worked, but any lengthy period during which it is stopped a certain amount of oxidation will occur; even a microscopic outgrowth of iron oxide (rust) forming will quickly wear out piston seals once the pump is started again. Obviously any thin internal coating or plating of a pump cylinder is not likely to last long due to wear. However, cast steel centrifugal pump casings are often quite satisfactory, although parts requiring critical clearances such as wear rings are usually inserts made of a more appropriate corrosion resistant metal. Similarly, cast steel centrifugal pump impellers are sometimes used; they are not of the same quality as non-ferrous ones, but are obviously a lot cheaper. Pumps with steel impellers usually cannot have close clearances and machined surface finishes, so their efficiencies are likely to be lower.

ii.   <u>Non-ferrous Metals</u>

Brass (a copper-zinc alloy) is commonly used for reciprocating pump cylinders. Due to its high cost, thin, seamless, brass tube is often used as a cylinder liner inside a steel casing, instead of a thick brass cylinder; obviously the steel must be kept from direct contact electrolytically with the water. Brass has good wear resistance in "rubbing" situations - i.e. with a leather piston seal, but it is not a particularly strong metal structurally, especially in tension. So called "Admiralty Brass" includes a few percent of tin, which greatly improves its corrosion resistance.

The bronzes and gun-metals are a large family of copper based alloys, which are generally expensive but effective in a wet environment; they usually have all the advantages of brass, but are structurally stronger, (and even more expensive too). Bronzes can contain copper alloyed with tin, plus some chromium or nickel in various grades and traces of other metals including manganese, iron and lead. So-called leaded bronzes replace some of the tin with lead to reduce costs, which still leaves them as a useful material for pump components. The inclusion of antimony, zinc and lead in various proportions produces the form of bronze known as gun-metal, which is a useful material for corrosion resistant stressed components. A bronze containing a trace of phosphorus, known as phosphor bronze, is an excellent material for plain bearings and thrust washers, if run with an oil film against a well-finished ferrous surface such as a machined shaft. Aluminium-bronze, which is cheaper but less corrosion resistant, replaces much or all of the scarce and expensive tin with aluminium.

Bronzes are generally among the best materials for making precision components that run in water and which need good tensile strength, such as pistons, valves, impellers, etc. Castings with a good finish can readily be obtained, and most bronzes machine very easily to give a precision surface.

Other materials, such as aluminium and the light alloys are generally not hard or wear-resistant enough for hydraulic duties, although by virtue of being very light they are sometimes used to make portable irrigation pipes; however they are not cheap as pipe material and can only be justified where the need to be able easily to move pipelines justifies the cost.

iii.   Timber

Timbers exist in a very wide variety of types; their densities can range from around 500kg/m³ (or less) up to 1,300kg/m³. They also offer a very large variation in mechanical properties, workability, wear resistance and behaviour in wet conditions. Timber is of course also susceptible to damage by insects, fungus or fire.

The most durable timbers are generally tropical hardwoods such as Greenheart, Iroko, Jarrah, Opepe, Teak and Wallaba. The durability of many timbers can be improved by treating them with various types of preservative; the most effective treatments involve pressure impregnation with either tar or water-based preservatives.

One of the main factors affecting the strength of a wooden member is whether knots are present at or near places of high stress. Where wood is used for stressed components, such as pump rods for windpumps or handpumps, it is important that it is finegrained and knot-free to avoid the risk of failures. Good quality hardwoods like this are not easily obtained in some countries and, where available, they are usually expensive. Cheap wood is limited in its usefulness and must be used for non-critical components.

Certain woods like lignum-vitae have also been used in the past as an excellent plain bearing material when oil lubricated running against a steel shaft, although various synthetic bearing materials are now more readily available and less expensive.

Wood is available processed into plywood and chipboards; with these a major consideration is the nature of the resins or adhesives used to bond the wood. Most are bonded with urea-based adhesives which are not adequately water resistant and are not suitable for outside use, but those bonded with phenolic resins may be suitable if applied correctly and adequately protected from water with paints. Therefore, for any irrigation device it is essential that nothing but "marine" quality plys and chipboards are used.

iv. Plastics

There is a large and growing family of plastics, which broadly include three main categories; thermoplastics, which soften with heat (and which can therefore readily be heated and worked, moulded or extruded); thermosets (which are heated once to form them during which an irreversible chemical process takes place) and finally various catalytically-cured resins. As with almost everything else, the better quality ones are more expensive. Great improvements are continuously being introduced and there are interesting composite plastics which contain a filler or a matrix of some other material to enhance their properties at no great cost.

Although plastics are weaker and softer than metals, they generally have the virtue of being compatible with water (corrosion is not a problem) and although their raw materials are not always low in cost, they do offer the possibility of low cost mass-production of pipe or components.

Thermoplastics based on polymerized petro-chemicals are generally the cheapest plastics; those used in the irrigation context include:

PVC (polyvinylchloride) is commonly used for extruded pipes; it can be rigid or plasticized (flexible); it is important to note that only certain grades of PVC (and other plastics) are suitable for pipes to convey drinking water for people or livestock, since traces of toxic plasticiser can be present in the water passed through some grades. PVC is relatively cheap and durable, but it is subject to attack by the UV (ultra-violet) wavelengths in sunlight and should therefore either be buried to protect it from the sun, or painted with a suitable finish to prevent penetration by UV radiation. PVC is also a thermoplastic and therefore softens significantly if heated above about 80°C; however this is not normally a problem in "wet" applications.

High density "polythene" (polyethylene) is cheaper and less brittle than PVC (especially at low temperatures) and is commonly used to make black flexible hose of use for irrigation, but it is also structurally much weaker than PVC, which is not necessarily a disadvantage for surface water conveyance at low pressure; however PVC is better for pressurized pipes.

Polypropylene is in the same family as polythene, but is intermediate in some respects in its properties between polythene and PVC. Polypropylene is less liable to fracture or to be sub-standard, due to bad management of the extrusion equipment, than is PVC; i.e. quality control is less stringent, so it can be more consistently reliable than poorly produced PVC.

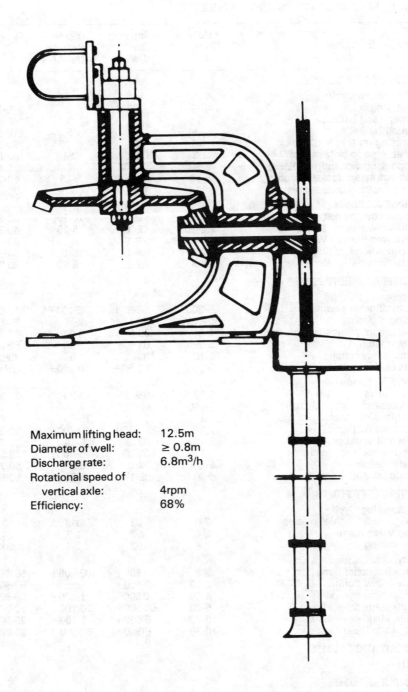

Fig. 79 Animal-powered Chinese Liberation Pump mechanism uses steel components (for strength) to good effect (see also Fig. 96).

Table 8 REVIEW OF PUMPS AND WATER LIFTS

Category and Name	Head Range (m)	Input Power (kW)	Flow Range (m³/h)	Efficiency (%)
I DIRECT LIFT DEVICES				
Reciprocating/Cyclic				
Watering can	5-3	.02	.5	5-15
Scoops and bailers	1	.04	8	40-60
Swing basket	.6	.06	5	10-15
Pivoting gutters & "Dhones"	.3-1	.04	5-10	20-50
Counterpoise or "Shadoof"	1-3	.02-.08	2-4	30-60
Rope & bucket and windlass	5-50	.04-.08	1	10-40
Self-emptying bucket, "Mohte"	5-10	.5-.6	5-15	10-20
Reciprocating bucket hoist	100+	100+	400+	70-80
Rotary/Continuous				
Continuous bucket pump	5-20	.2-2	10-100	60-80
Persian wheel or "tablia"	1.5-10	.2-.6	5-25	40-70
Improved Persian W. "Zawaffa"	.75-10	.2-1	10-140	60-80
Scoop wheels or "Sakia"	.2-2	.2-1	15-160	60-80
Waterwheels or "Noria"	.5-8	.2-1	5-50	20-30
II DISPLACEMENT PUMPS				
Reciprocating/Cyclic				
Piston/Bucket pumps	5-200+	.03-50+	2-100+	40-85
Plunger pumps	40-400	.50-50+	2-50+	60-85
Diaphragm pumps/IRRI pump	1-2	.03-5	2-20	20-30
"Petropump"	5-50	.03-5	2-20	50-80
Semi-rotary pumps	1-10	.03-.1	1-5	30-60
Gas or Vapour displacement	5-20	1-50+	40-400+	n/a
Rotary/Continuous				
Flexible vane pumps	5-10	.05-.5	2-20	25-50
Progressive cavity (Mono)	10-100	.5-10	2-100+	30-70
Archimedean screw	.2-1	.04	15-30	30-60
Open screw pumps	2-6	1-50+	40-400+	60-80
Coil and spiral pumps	2-10	.03-.3	2-10	60-70
Flash-wheels & Treadmills	.2-1	.02-20	5-400+	20-50
Water-ladders	5-1	.02-1	5-20	50-70
Chain (or Rope) and Washer	5-20	.02-1	5-30	50-80
III VELOCITY PUMPS				
Reciprocating/Cyclic				
Inertia and "Joggle" pumps	2-6	.03	1-3	20
Flap valve pump	2-6	.03	1-3	20
Resonating joggle pump	2-6	.03	2-4	50
Rotary/Continuous				
Propeller (axial flow)	5-3	10-500+	100-500+	50-95%
Mixed flow pumps	2-10	150-500+		50-90%
Centrifugal pumps	4-60	.1-500+	1-500+	30-80%
Multi-stage mixed flow	6-20	50-500+	10-100	50-80%
Multi-stage centrifugal	10-300	5-500+	1-100	30-80%
Jet pump Centrifugals	10-30	5-500+	50-500	20-60%
IV BUOYANCY PUMPS				
Air lift	5-20			
V IMPULSE PUMPS				
Hydraulic ram	10-100			
VI GRAVITY DEVICES				
Syphons, Qanats or Foggara	1-6			

None of the above plastics are generally applicable for manufacturing pump components for which strength and durability are important; these require more expensive and specialized plastics, such as nylons, polyacetals and polycarbonates. Nylons can be filled with glass, (for strength), molybdenum disulphide (for low friction), etc. An expensive specialized plastic of great value for bearings and rubbing surfaces on account of its low friction and good wear resistance is PTFE (polytetrafluoroethylene); certain water lubricated plain bearings rely on a thin layer of PTFE for their rubbing surface, and this proves both low in cost and extremely effective.

There are also various specialized thermoset plastics which find applications as pump components; these tend to be tougher, more wear resistant and more heat resistant than thermoplastics, and therefore are sometimes used as bearings, pump impellers or for pump casings. They are also useful for electrical components which may get hot. Most "pure" plastics are inclined to creep if permanently loaded; i.e. they gradually deform over a long period of time; this can be avoided and considerable extra strength can be gained, through the use of composite materials where glass fibre mat (for example) is moulded into a plastic. Various polyesters and epoxides are commonly used to make glass-reinforced plastics (g.r.p. or "fibreglass"); these are used to make small tough components or, in some cases, to make large tanks. Another example of composite plastics is the phenolic composites where cloth and phenolic resin are combined to make a very tough and wear resistant, but readily machinable material which makes an excellent (but expensive) water lubricated bearing, such as "Tufnol".

3.13 SUMMARY REVIEW OF WATER LIFTING DEVICES

Table 5, which introduced this chapter, is sorted into categories of pump types based on their working principles, but it is difficult to see any pattern when looking through it. Therefore, Table 8, which concludes this chapter attempts to quantify the characteristics of all pumps and water lifts in terms of their operating heads, power requirements, output and efficiency. Finally, Fig. 80 (A, B and C) indicates the different categories of pump and water lift demarcated on a log-log head-discharge graph (similar to that of Fig. 11). Obviously there are no hard and fast boundaries which dictate the choice of pump, but the figure gives a graphic indication of which pumps fit where in terms of head and flow, and hence of power. Note that Table 8 shows input power requirements, whereas Fig. 80 gives the hydraulic power produced, which will be a lesser figure by the factor of the pump efficiency. Due to the use of the log-log scales, the smaller devices appear to occupy a larger area then they would if linear scales had been chosen, however in this case it would not have been possible to fit sufficient detail in to the corner where the multiplicity of low-powered, low-head and low-flow devices fit, had a linear scale been applied.

4. POWER FOR PUMPING

Fig. 81 indicates the most feasible linkages between different energy resources and prime movers. It shows how all energy sources of relevance to small or medium scale irrigation pumping originate from renewable energy resources or from fossil fuels; the arrows then show all the routes that can apply from an energy resource to produce pumped water. In some cases similar components can be used within systems energized in completely different ways; for example electric motors are necessary either with a solar photovoltaic pumping system or with a mains electrical system, so the motor-pump sub-systems of both types of system can have a lot in common.

The details of the components in Fig. 81 are discussed through the following section, but it is first worth reviewing a few generalities relating to the combination of prime movers and pumps.

4.1 PRIME MOVERS AS PART OF A PUMPING SYSTEM

4.1.1 Importance of "Cost-Effectiveness"

Almost every aspect of an irrigation pumping system consists of compromises, or trade-offs, between the capital (or first) cost of the system and the running (or recurrent) costs. Farmers tend to purchase cheaper systems with higher running costs, resulting in the widespread use of other than the most efficient and cost-effective systems. No doubt this is because farmers are generally short of capital and often in any case regard large capital investments as being inherently more risky than incurring regular running costs, (which in time may mount up to a large sum).

There is therefore a good case for institutional users with an interest in improving agricultural techniques, such as agricultural credit agencies, aid agencies and governments, to try to assist in overcoming these problems by providing suitable financial inducements to encourage the use of more cost-effective irrigation systems.

In the final analysis, the combination of components to make a pumping system will depend on the cost effectiveness of the total chosen system under whatever specific technical, agricultural and financial conditions happen to be prevalent. That is "cost-effectiveness" in the very broadest sense, including not just the first costs and running costs for the system, but factors like the convenience and ease of use as perceived by the farmer; (i.e. including, in economists' jargon, the "opportunity costs"). In reality, the selection decision is usually limited to what is known to be available and affordable and yet is capable of the required pumping duty.

The selection process is discussed in more detail in the next section, but certain considerations inherent in the economics and hence in the relative cost-effectiveness of different system choices are important in relation to any discussion of linking components to make up pumping systems. To this end, some of the cost attributes of different prime-mover options are compared in Table 9. Here the main categories of irrigation pumping system are reviewed

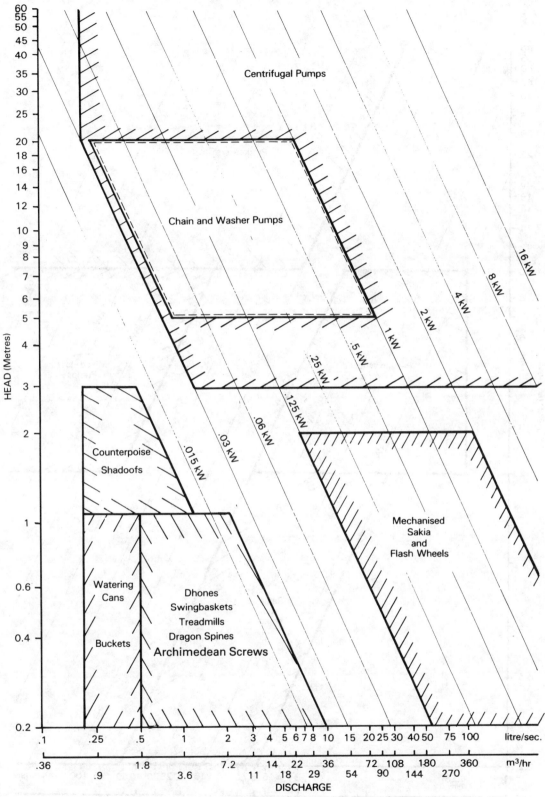

Fig. 80a Typical head and discharge capacities for different types of pumps and water-lifting devices (on a log-log scale) (and continued in Figures 80b and 80c)

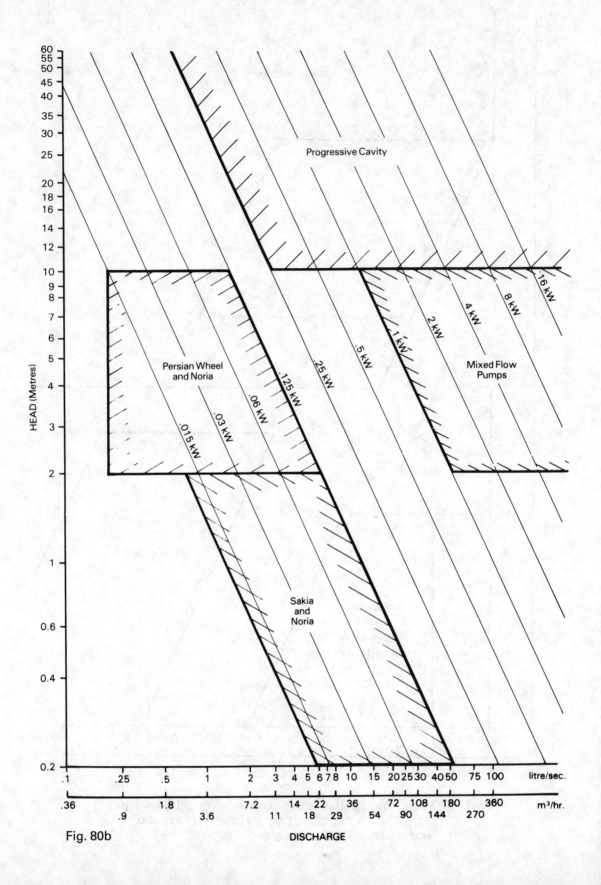

Fig. 80b DISCHARGE

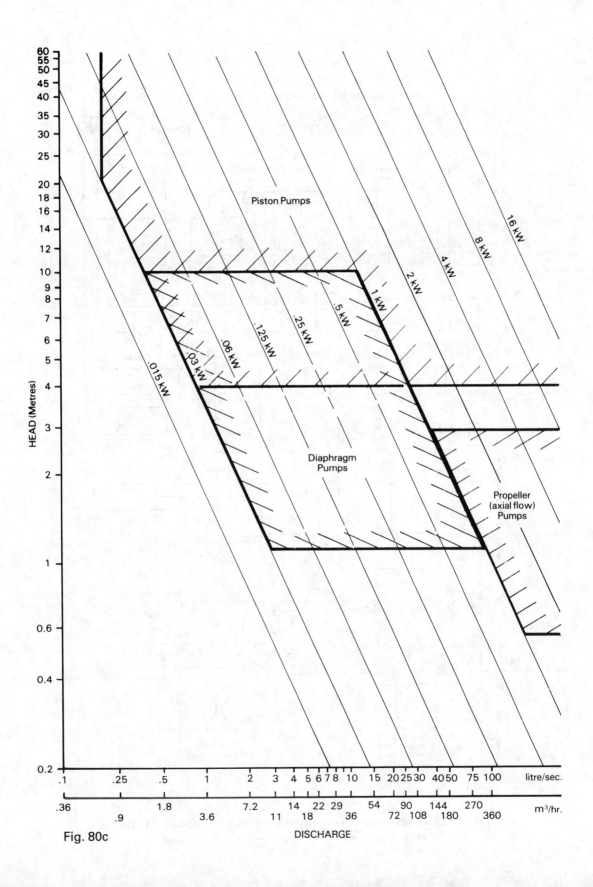

Fig. 80c

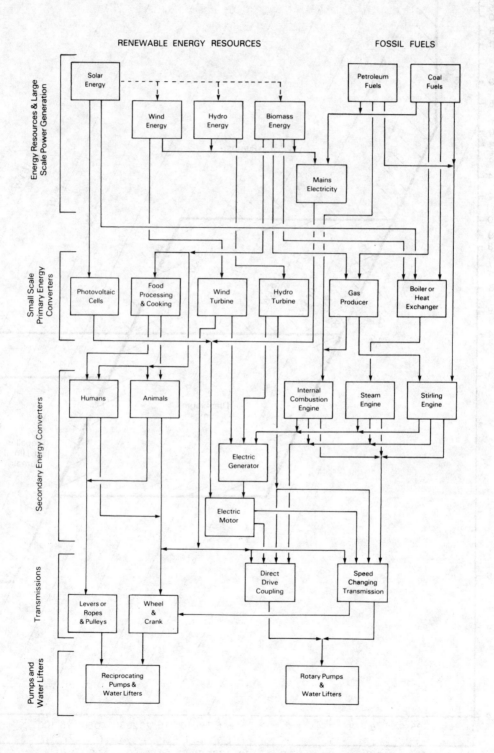

Fig. 81 Linkages between energy resources and appropriate prime movers

Table 9   COST ATTRIBUTES OF PRIME MOVERS

Cost	Prime mover	Human	Animal	Solar	Wind	Hydro	Biomass fuelled engines	Mains[1] electricity	Petrol-fuelled engines	Coal fuelled engines
First Costs	Capital	*	**	*****	****	****	****	* or *****	**	****
	Shipping	*	*	***	****	***	****	**	**	****
	Installation	*	**	*	***	***	***	* or *****	**	***
Recurrent Costs	Fuel	*****	***	NIL	NIL	NIL	**	***	***	**
	Spares	*	*	*	*	*	****	*	****	***
	Maintenance	*	*	*	*	*	****	**	****	*****
	Attendance[2]	*****	****	*	*	*	***	*	***	***
Productivity		*	**	** to ***	** to ****	****	****	****	****	****
Opportunity Costs		*****	***	*	*	*	**	***	****	**

Key   * = Low
      ** = Low to Moderate
      *** = Moderate
      **** = Moderate to High
      ***** = High

[1] Electricity can vary from low to high depending on whether a mains connection is already available or not.
[2] "Attendance" implies level of human intervention needed.

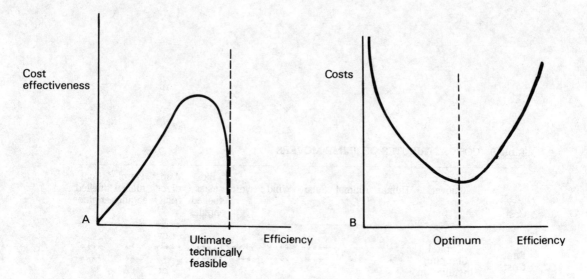

Fig. 82 A. How diminishing returns eventually defeat the benefits of seeking increased efficiency beyond certain levels
B. The influence of efficiency on costs

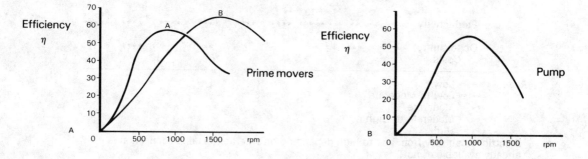

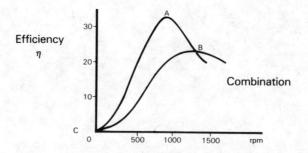

Fig. 83 Illustration of how correct speedmatching of a prime mover to a pump can be more important than the efficiency of the prime mover. Here prime mover 'A' is less efficient than 'B' but has a better speed match to the pump — hence the less efficient prime mover 'A' provides a more efficient system

in terms of first costs, recurrent costs and then in terms of two factors relating more to their "effectiveness" (i.e. their productivity and their general availability). It is clear from this table that no single method offers both low first costs and low recurrent costs and yet is also among the most productive (otherwise it would be universally applicable and the other options would be of little interest). High productivity depends either on relatively high first costs (when using renewable energy equipment generally) or on relatively high running costs combined with moderately high first costs when using fossil-fuelled equipment. In the end, successful selection depends on the choice of the best trade-off between the availability of finance, the capability of maintaining and financing the recurrent costs of the system and the performance or productivity that is expected.

It is possible that potentially useful systems are usually not considered simply because, not being conventionally used at present, they are unfamiliar and therefore are not known or understood well enough by potential buyers; this is where it is hoped that publications such as this may encourage some attempts to try new methods, preferably by institutions or individuals with the resources to underwrite the risks inherent in experimenting with new or unfamiliar technologies.

i. <u>Cost-effectiveness and efficiency</u>

Generally, a cost-effective system needs to be technically efficient; i.e. a relatively high output is needed in relation to the energy input. This is just as true for renewable energy powered systems as for fossil fuelled systems. In the former case, the energy resource, if it is solar energy, wind or water power is notionally cost-free, but the capital cost of the system is closely linked to the efficiency. This is because for a given pumping requirement, if you halve the efficiency of the system you must double the "cross section" of the energy resource to be intercepted; i.e. you need twice the area of solar collector or twice the rotor area of a windmill, or a turbine capable of passing twice the flow rate of water. This tends to require a system that is twice as large and therefore usually twice as expensive.

In all cases there is an ultimate technical efficiency that can be approached but never quite achieved, (Fig. 82 A). Pursuing the cause of better efficiency is usually worthwhile up to a point, but thereafter it brings diminishing returns as increasing complication, sophistication and cost is required to achieve small further gains in efficiency. However it usually requires a mature technology to be at the level where further improbvements in efficiency are counter-productive, and in any case, new manufacturing processes and materials or increases in recurrent costs (due to inflation) sometimes allow improvements to become cost-effective in the future which were not justifiable in the past.

The influence of efficiency on costs is illustrated in Fig 82 B, which shows how low efficiency generally causes high costs and that there is an optimum range of efficiency for most technologies where reasonably low costs are achieved, but above which diminishing returns set in. In the case of renewable energy systems these costs will be largely attributable to the capital cost and hence to financing the investment, while in the case of fossil fuelled devices a large proportion of the costs will relate to running and maintaining the system.

ii. **Combining system components with differing efficiencies**

Virtually all pumping system components achieve an optimum efficiency at a certain speed of operation. Some components like pipes and transmission systems are most efficient (in terms of minimizing friction and hence losses) at very low rates of throughput, but they are then least productive and they will therefore have a point of "optimum cost-effectiveness" where there is a good compromise between their productivity and their efficiency. Prime movers invariably have an optimum speed of operation; this is as true of humans and animals as it is of diesel engines or windmills.

Fig. 83 shows three sets of curves; first, efficiency against speed for two prime-movers, (in this example electric motors would fit the speeds and efficiencies shown); second, a curve for a typical pump and lastly, for the combination of the prime-movers with the pump. It must be remembered that the efficiency of a combination of two components is numerically the product (i.e. the multiplied result) of their individual efficiencies; eg. a 30% efficient engine (0.3) with a 50% efficient pump (0.5) has a combined efficiency of:

$$0.3 \times 0.5 = 0.15 \quad \text{or} \quad 15\%$$

The important point contrived in Fig. 83 is that the prime-mover with the highest optimum efficiency is not in this case the best one to use with a particular pump. In the example motor "A" has a best efficiency of 58% while motor "B" achieves 66%, yet, because the optimum efficiency of motor "A" occurs at a speed which coincides well with the optimum efficiency of the pump, the combined efficiency of that combination is better if the motor is direct-coupled to the pump; (motor "B" will drive the pump at a speed greater than its optimum, as at 1,500 rpm the pump has an efficiency of only 35%) so the best efficiencies of the two alternative combinations are:

1,000 rpm motor and pump $0.58 \times 0.55 = .32$ or 32%

1,500 rpm motor and pump $0.66 \times 0.35 = .23$ or 23%

This illustrates how it is generally more important to ensure that the design speeds of components match properly than to ensure that each component has the highest possible peak efficiency.

4.1.2 Transmission Systems

Components often do not match effectively; i.e. their optimum speeds of operation are different. In such situations it generally pays, and it is sometimes essential, to introduce a speed changing transmission. Also, in many situations the prime-mover cannot readily be close to the pump, and some method is therefore necessary for transmitting its output either horizontally or vertically to the water lifting device.

i. **Transmission principles**

Power can be transmitted from a prime mover to a pump in a number of ways; the most common is a mechanical connection, which can either rotate (shafts, belts or gears) or reciprocate (pump rods or levers). Where power has to be transmitted some distance, then electricity, hydraulic pressure or compressed air can be used, since it is difficult to transmit mechanical power any distance, especially if changes of direction or bends are needed.

In all transmissions there is a trade-off between the force or torque being transmitted by the system (which demands robustness to resist it) and the speed of operation (which tends to cause wear and reduced life). Power, which is what is being transmitted, can be defined as the product of force and velocity. Mechanical systems that run at slow rotational or reciprocating speeds need larger forces to transmit a given amount of power, which in turn require large gear teeth, large belts or large pump rods (for example) and these inevitably cost more than smaller equivalents. Where mechanical power is transmitted some distance any reciprocating linkages need to be securely anchored; (even a 5m farm windpump can pull with a reciprocating force peaking at about 1 tonne). For this reason, most modern commercial systems involving lengthy mechanical links tend to use high speed drive shafts (for example surface mounted electric motors driving a rotodynamic pump located below the water (or below flood level) as in Figs. 66 or 134 B). A high speed drive shaft can be quite small in section because its high speed results in low torque. However a high speed drive needs to be built with some precision and to have good (and expensive) bearings to carry it and to align it accurately so as to prevent vibrations, whirling of the shaft, premature wear and other such problems.

Electrical, hydraulic or pneumatic transmissions all have a common requirement demanding that their voltage, or pressure of operation ideally needs to be high to minimize the cross section of cable or of pipe needed to transmit a given power flow efficiently. High voltage cable (or high pressure pipes), need to be of a good quality and inevitably cost more per metre for a given cross section. Therefore, with all transmissions there is a trade-off between efficiency and cost; cheap transmissions often reduce the capital costs but result in high recurrent costs due to their lower efficiency and greater maintenance and replacement needs, and vice-versa. It is therefore advantageous to match prime-movers and pumps of similar speeds to avoid the cost and complication of speed-changing transmissions.

ii.   Mechanical transmissions

The most common need for a mechanical transmission is to link an engine or an electric motor with a pump. Generally such prime-movers are used with centrifugal or other rotodynamic pumps which run at the same speed as the engine or motor; in such situations they can be direct-coupled with a simple flexible drive coupling as in Figs. 72 and 105. Speed changing of up to about 4:1 can readily be achieved with vee-belts as shown in Figs. 84, 99 and 106. Fig. 84 shows a two stage vee-belt drive where the total speed change can be as much as 4:1 on each stage. In this situation the total speed change is the product of the ratios for each stage. Where multiple vee belts are needed on one drive stage, as in Fig. 99 (showing four in use), it is best to use matched sets from a supplier, and always to renew all belts simultaneously so that they all share the load effectively; a more modern and convenient type of belt is the so-called poly-vee, which is similar to a whole lot of small vee-belts fixed together edge-to-edge. Flat belts (made of leather) used to be common and they are coming back, sometimes today made of synthetic materials, as they are more efficient with less friction than a set of vee belts.

If a speed change greater than about 4 or 5 to 1 is needed, then an alternative to multiple stages of belts (which introduce problems with belt adjustment) is to use gearboxes. A right angle drive may be created to drive a vertical shaft borehole pump (for example) either by using a 90° geared well head, or by using a twisted flat belt. To be successful twisted belt drives need to have a generous distance between the pulleys in relation to their diameters or excessive wear will occur.

Other mechanical transmissions commonly used are reduction gearboxes with a pitman drive, similar in most respects to the windpump transmission of Fig. 109. They consist of a rotary drive shaft which drives a single or pair of larger gear wheels via a small pinion; the large gear wheels drive a reciprocating cross-head or pitman slider via two connecting rods. The pump rod is connected to the cross-head or pitman. Mechanisms of this kind can be used to connect a diesel engine or an electric motor to a reciprocating piston pump. Other mechanical right-angle drives are illustrated by reference to Figs. 94, 95 and 96, (the large size necessary for making a strong enough drive from traditional materials is well-illustrated in Fig. 95).

When budgeting for a pumping system, it is important to know that the mechanical transmission can cost as much or often more than the prime-mover, especially if a geared or reciprocating well-head is used. The high cost is due to the mechanical requirements for reliable operation being demanding and the volume of production usually being much lower than for engines or electric motors.

An effective method of transmitting mechanical shaft power any distance is via a high speed rotating shaft. This needs to be steadied by bearings at quite close intervals to prevent the shaft "whirling" like a skipping rope, a phenomenon which causes intense vibration and destruction of the shaft. Vertical drive shafts down boreholes, as much as 100m deep and running at 1,500rpm or more are commonly and successfully used, although submersible electric multi-stage pumps are becoming a more popular solution.

iii.   Electrical, hydraulic or pneumatic transmission

The use of a diesel-generating set (or wind-electric, solar-electric or hydro-electric unit) as a prime mover allows considerable flexibility in transmission (literally) since electric cable is all that is needed to link the prime mover to a motor-pump unit, (which can even be submerged down a borehole as in Figs. 71 or 134 A).

Other options, which are technically feasible, but more rarely used are hydraulic or pneumatic transmissions in which either a liquid (water or oil) or air are pumped through pipes to drive a pump. Examples of hydraulic transmissions are given with the jet pump in Fig. 75, or the positive displacement hydraulically activated pumps of Fig. 39. The air lift pump of Fig. 76 is an example of a pumping system which requires pneumatic transmission. Pneumatic diaphragm pumps are commercially available and tend to be most commonly used for construction projects, with an air supply from mobile engine driven air-compressors. They are not normally used for on-farm irrigation but there is no technical reason why they would be unsuitable. However, hydraulic and pneumatic transmissions tend to be inefficient and therefore such a system may have high running costs.

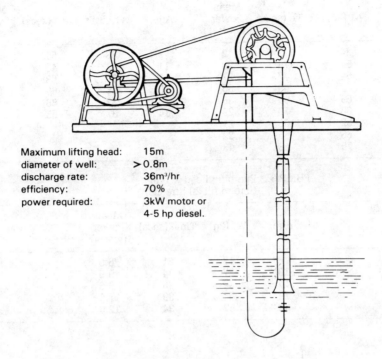

Fig. 84 Two stage speed reduction transmisson used in China to connect an electric motor to a chain and washer pump

Table 10 THE CALORIFIC VALUES OF VARIOUS STAPLE FOODS

Staple Crop	Energy Content MJ/kg	Kg/day to provide 10MJ
Dayak rice	10.4-11.4	0.92
Iban rice	13.3	0.75
Tanzania rice	8.2	1.22
Maize (Africa)	4.2	2.38
Millet (Africa)	3.8	2.63
Sweet potato (Africa)	10.1	0.99
Cassava (Africa)	15.0	0.67
Yams (Africa)	9.5	1.05
Groundnuts (Africa)	7.2	1.39

After Leech, reference [6].

Table 11 POWER CAPABILITY OF HUMAN BEINGS

Age	Human power by duration of effort (in watts)					
Years	5 min	10 min	15 min	30 min	60 min	180 min
20	220	210	200	180	160	90
35	210	200	180	160	135	75
60	180	160	150	130	110	60

PUMP 1.	Pumping head 10.54m time to fill 20 litre can		
Age	Wt (kg)	Time (secs)	Mean Power (watts)
9	34	173	12
14	54	74	28
14	54	77	27
16	50	69	30
18	55	59	35
20	68	70	29
29	82	100	21
33	65	55	32
47	75	48	43

PUMP 2.	Pumping head 6.35m Time to fill 20 litre can		
Age	Wt (kg)	Time (secs)	Mean Power (watts)
9	33	51	29
11	31	48	28
11	31	68	18
10	55	60	20
10	32	66	19
14	37	55	22

PUMP 3.	Pumping head 2.14m time to fill 20 litre can		
Age	Wt (kg)	Time (secs)	Mean Power (watts)
6	17	61	6.9
10	27	44	9.5
14	64	36	6.6
19	54	39	11.0
38	57	34	12.0

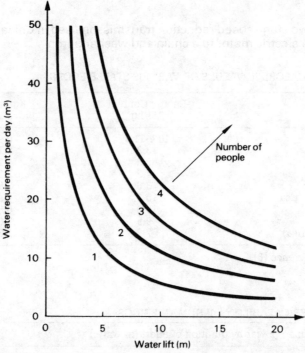

Fig. 85 The number of people required to provide a specified quantity of water at different lifts. The curves have been derived by assuming that a single person can provide 60 watts of power for 4 hours per day and that the efficiency of the pump is 60%

4.1.3 Fuels and Energy Storage

Power sources need energy, whether it is fuel for an engine, wind for a windpump or sunshine for a solar pump. The main difference is that the provision of fuel can usually be arranged by the user, but nobody can make the wind blow or make the sun shine on demand. There is therefore an obvious qualitative difference between wind and solar powered devices which will only function under certain weather conditions and the rest which generally can be made to operate at any pre-planned time.

Although the apparent randomness of wind or solar availability would appear to be a serious disadvantage, in reality the energy available over a period of a few days in a given location at a given time of the year does not vary much from year to year. The problem is more one of covering a mismatch that can occur between the rate at which energy is available and the duty cycle the farmer would like to impose. This can often be overcome either by choice of technique or by including a storage facility.

In most cases, where the output required is water, the most cost-effective solution is to introduce a storage tank between the pump and the field; (in some cases the field itself can act as a storage tank). The other principal method for small scale energy storage is to use lead-acid electrical batteries, but this becomes prohibitively expensive except when small amounts of energy of less than about 1-2kWh need to be stored. The costs of tanks for storing water relate to their volume, while the costs of batteries relate to their energy capacity; therefore, at low heads when large volumes of water may need to be stored, but which involve little energy, electrical battery storage can be cheaper (and less demanding in terms of land utilisation) than storage tanks. However, before considering the substitution of batteries for storage tanks, it must be remembered that batteries would need replacing a lot more often than the storage tank and would also need much more maintenance.

4.2   HUMAN POWER

4.2.1 Human Beings as Power Sources

In the whole small-scale pumping field it is generally difficult to make precise statements on pumping performance which are generally correct; nowhere is this more true than in the field of human powered waterlifting devices and pumps. This is partly because human capabilities are very variable, but also because there is a multiplicity of pumps and water lifts of widely varying efficiency.

i.   Efficiency as prime movers

People (and animals) derive their power from the calorific content of their food. Even when physically inactive the human body requires energy to run its basic metabolic functions, i.e. to power the heart and circulate blood, to work the lungs and digestive system, etc. Energy for muscle power is then an extra requirement on top of this. A typical food energy requirement is around 2400kcal, 10MJ or 2.8kWh per 24 hrs. Table 10 indicates the calorific values of various staple foods (after Leech [6]).

A person's muscular work capability per day is in the region of

200-300Wh/day. Human beings therefore have an average overall efficiency in the region of 7-11% for converting food energy to mechanical "shaft energy". This figure includes the basic metabolic energy requirement; the efficiency of the muscles for short but strenuous efforts can be as high as 20 or 30% [12] & [19], which compares well with internal combustion engines.

Kraatz, [12] (quoting Wood), gives a calculation of the food required by a man to generate the energy needed to irrigate a crop. On the assumptions of a rice crop needing 850mm of water in 120 days, with a yield of 600kg of rice from a 0.2ha plot, with a 50% efficient water lifting device lifting the necessary water through 3m head, the marginal cost of "fuelling" the human prime-mover for the irrigation pump was calculated to be 35kg of rice or 6% of the expected total yield. An additional 35kg would be needed to cover the basic metabolism of the person concerned, giving a total of 70kg or 12% of the rice produced.

The Intermediate Technology Development Group's Water Panel [20] gave a rule of thumb of a food requirement of 0.5kg of rice per MJ of hydraulic work, plus 0.012kg of rice per day per kg body weight. In the earlier example, the hydraulic requirement was 50MJ, which under the rule of thumb just quoted demands 25kg of rice for pumping effort, and a 60kg man would additionally need (0.012 x 60 x 120 = 86kg) of rice, giving a total requirement of 111kg of rice, or 18% of the total crop produced.

Allowing for losses of rice, possible worse yields than that assumed, and the food requirements of the farmer's dependants, it is easy to see how hard it is to generate a surplus when cultivating staple crops on small land holdings. For example, if he has three dependants and loses just 20% of his crop through various forms of wastage, the farmer and his family will need to retain 60 to 90% of the harvest, depending on the method of estimating rice requirements used. Slightly worse wastage or a larger family would result in barely sufficient food for pure subsistance.

ii   Productivity

Contrary to popular belief, human muscular energy is not cheap. The poor are forced to use human power, usually because they cannot afford anything better, since the cash investment required is minimized and therefore it is more "affordable" than other options. As will be shown, almost any other source of power will pump water more cheaply unless only very small quantities are required.

The human work capability is around 250Wh/day, <u>so it takes four days' of hard labour to deliver only one kWh</u> - which a small engine could deliver in less than one hour while burning less than one litre of petroleum fuel. So the farmer with a small mechanized pumping system has the equivalent of a gang of 20 to 40 men who will work for a "wage" or running cost equivalent to say 1 litre of fuel per hour; not surprisingly, any farmer who can afford it will sooner choose to employ an engine rather than 20 to 40 men. This argument can be turned on its head to show the high price of human muscle power, if the "opportunity cost" or a real wage cost is assigned to human muscular labour; eg. assuming a daily wage rate of US $1.00/day gives an energy cost of about $4.00/kWh. Although this is a low wage for hard labour, even in some of the poorer countries, it represents an energy cost that is significantly more expensive than even new and exotic power sources such as solar photovoltaic panels.

There is an opportunity cost caused by diverting people from more important work to pumping water; the best asset people have is brains rather than muscle; therefore, if agricultural productivity is to improve and economic standards are to be advanced, it is essential to introduce more productive power sources for all except the very smallest of land-holdings.

iii. <u>Power capability</u>

Muscle power can handle quite large "overloads" for short periods, but the power capability diminishes if more than a few minutes of activity are required. The power availability is also a function of the build, age, state of health and weight of an individual; finally the ability to produce power depends on the nature of the device being worked and the muscles that can readily be utilized. Table 11 A indicates power outputs that may be expected for individuals of 20, 35 and 60 years age respectively over periods of operation ranging from 5 minutes to 3 hours, after Hofkes [21], presumably with devices allowing much of the body to operate. Table 11 B shows actual results measured by the Blair Research Institute in Zimbabwe, [22].

Therefore, although the actual output from any human powered pump is not precisely predictable, an approximate prediction can reasonably be made. Fig. 85 gives a set of curves indicating the capability of from one to four people each providing 240Wh per day of useful work through a pumping device with an efficiency of 60%. These curves probably represent what is generally achievable under favourable circumstances and indicate, for example, that the daily output per person if lifting water 5m, is about 12m³.

iv. <u>Ergonomics</u>

The actual useful output from a person depends a lot on the way the water lift or pump works; the most powerful muscles are the leg and back muscles while the arm muscles are relatively weak, so conventional hand pumps are less effective at "extracting work" from a person then a device like a bicycle. Moreover, the "ergonomics" of the design are important; the operator needs to be comfortable and not contorted into some difficult position, so the device should require a relaxed posture, with the user well-balanced, and it should function best at a comfortable speed of operation. Utilization of the leg muscles will also often allow the operator to throw his or her weight behind the effort in order to gain further pedal pressure. Wilson [23] reported that a rotary hand pump was improved in output by a factor of three, (300%), by converting it from hand operation to foot operation. The same article also promoted the bicycle as a supreme example of effective ergonomics; it uses the right muscles in the right motion at the right speed and applies human power through a light but strong and efficient mechanism. Wilson makes the very valid point that what is needed is a pump which is as well designed, strong, efficient and easy to use as the bicycle. He quotes dynamometer tests as indicating that the average cyclist works at 75W when cycling at 18km/h; if this output could be produced while pumping, the following flowrates should be realisable at various lifts, assuming a water lifting device of only 50% efficiency:

Head:	0.5	1.0	2.5	5.0	10.0	m
Flow:	27.5	13.8	5.5	2.2	1.1	m³/h

For this reason the most effective irrigation pumps are in fact foot operated. Also an irrigation pump requires to be operated perhaps for several hours and therefore efficiency and ease of use are crucial. Hand operated devices are easier to install and can be lighter and smaller (since no one has to stand or sit on them and the forces that can be applied will not be so great anyway). Where pumps are used for water supply duties rather than irrigation, efficiency is less of a stringent requirement since any individual user will generally only operate the pump for a few minutes per day to fill a few small containers.

Therefore the criteria for defining a good human powered irrigation pump are significantly different for those for a water supply pump and it may be a mistake to use pumps for irrigation duties that have only been proved successful in water supply.

4.2.2 Traditional Water Lifting Devices

Many of traditional water lifting devices are particularly designed for low lift irrigation, and they are often foot-operated since it no doubt became apparent that this was the best method of harnessing human power.

The least-cost solution has always been a bucket or bag of water lifted when necessary on a rope; Fig. 86. The best that can be said for this technique is that with small plots the water can at least be applied with precision to individual plants, so at least efficient conveyance and distribution can partially compensate for the inefficiency of the actual water lifting. At low heads, the use of buckets and scoops, (see also section 3.3.1) led to the development of the swing-basket, (Fig. 18), which can use two people and functions more rapidly, although only through very low pumping heads such as from canals into paddies. However, it is not an ergonomic device in that a lot of muscular effort goes into twisting the body, there is much spillage, and also water is lifted much higher than necessary. Nevertheless, two young boys using this technique, for example in Bangladesh, can complete 2,000 swings without a rest, according to Schioler [91].

An improvement, obtained at the price of some slight complexity, is the use of suspended or pivoted devices such as the supported scoop (Fig. 19) and some which are also balanced such as the dhone (or dhoon) see-sawing gutters, (Fig. 20) or the counterpoise lift (or shadoof), (Fig. 21). These are no longer portable since they need to be installed on a site, and they require a supporting structure which has to be attached securely to the ground, but they are far more efficient than such primitive devices such as buckets or swing baskets, as indicated by the performance curves taken from Khan, [25], in Fig. 87. This shows that a single dhone will lift 7.5 litre/sec at a lift of 0.75m (115 gall/min at 30 inches), which reduces to about 2 litre/sec (30 gall/min) at 1.5m (or 60 inches) head. Therefore the dhone will move more than twice as much water as a swing basket at low lifts, moreover using the power of only one person rather than two. Khan makes the point that many Bangladesh farmers try and use a single stage dhone at too high a lift, and lose a lot of performance as a result; the optimum lift per stage is approximately 1m (40in).

Table 12, adapted from Khan, [25], indicates how widely the dhone is used in Bangladesh, a country with very large areas offering the possibility of shallow lift irrigation, and it also shows clearly how much of an improvement the dhone is over the swingbasket. The same table also indicates

Fig. 86 Rope and bag water lift from a dug well (example from The Gambia)

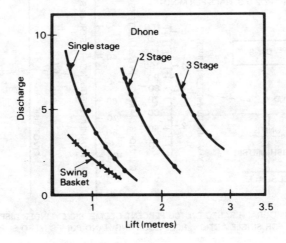

Fig. 87 Relative performance of the swing basket and the Dhone (after Khan [25])

Table 12 COMPARISON OF VARIOUS WATER LIFTS IN BANGLADESH[1]

	Dhone	Swing	Dugwell[2]	Mosti[3]
Area irrigated				
ha × 1000	392	65	4	20
percentage of total	35	6	0.8	
Water source	Surface	Surface	Pumped	Pumped
Max discharge m³/hr	7.5	2.3	0.6	0.8
Pumping head m	0-1.5	0-1.8	0-4.5	1.5-6.0
Capital cost US$	20.00	1.33	10.00	80.00
Working life years	4	2	3	6
Command area				
ha of dry season paddy	1.6-2.0	.4-.6	0.3	0.2

[1] Adapted from Khan, [25]
[2] Hand dug well with counterpoise bucket lift
[3] Manually operated shallow tubewell for irrigation
 Uses industrially made headpump (No. 6. Carbiron pump)

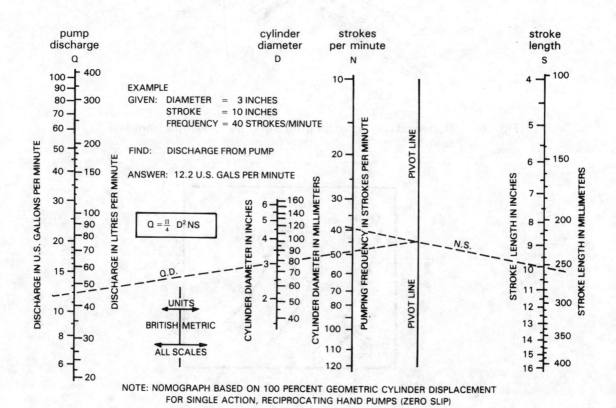

Fig. 88 Nomograph for calculating hand pump discharge (after Hofkes [21] and McJunkin [26]).

the characteristics relating to the "dugwell", a counterpoise lift and shallow hand-dug well, and the "Mosti", which is a cast iron handpump mounted on a tubewell.

Despite being a considerable improvement on the swingbasket, the dhone is probably not as efficient as the various rotary devices, described in Section 3.6, although it represents a good compromise between retaining simplicity and low-cost while achieving a useful output.

Rotary devices tend to be easier to work as they generate a smooth output and they therefore often can be driven with a comfortable pedaling motion of either the arm or, better, the legs. The various flash-wheels and ladder or dragon spine pumps are generally leg-powered, (see Sections 3.6.5 to 3.6.7 plus Figs. 51 and 52), and they can often readily be powered in this way by several operators. While today Archimedean screws are usually hand operated (Fig. 48), in Roman times they were walked on rather like the treadwheel in Fig. 51, which no doubt was easier for the operator and more productive.

It is quite possible that further useful improvements could be made with some of the simple traditional water lifts. This is an area where study and experimentation, perhaps as part of technical educational programmes or by NGOs working in the field, might yield useful results.

In many cases human powered devices offer the best means to initiate small scale lift irrigation because of their low first cost. However, in the longer term it is to be hoped that small farmers will be assisted to advance towards more productive pumping techniques, which inevitably require some mechanization.

4.2.3 Handpumps

Handpumps, although less productive than footpumps, are the most common form of industrially manufactured manually operated water lift, and for that reason are very widely used. The classic design of piston pump or bucket pump is shown in Figs. 29 and 30. Most of these pumps were developed for use by a family to provide water for themselves and their livestock, rather than for irrigation. The problem when pumps of this kind are used for irrigation is the intensity with which they are used compared with their use for water supply; instead of pumping for a few minutes per day they have to be used for several hours per day, which naturally tends to shorten their useful lives considerably and also to increase the incidence of breakages.

The forces involved in driving a piston pump with a lever have already been discussed in Section 3.5.1. So far as handpumps are concerned, Hofkes [21] indicates the following maximum heads as being generally suitable for comfortable operation of various common sizes of handpump:

cylinder diameter		maximum head	
(mm)	(in)	(m)	(ft)
51	2	25	75
63	2½	20	60
76	3	15	45
102	4	10	30

The load at any given head can be reduced by shortening the stroke of the pump, but the above recommendation presumably applies with typical strokes in the region of 150 to 300mm.

Hofkes [21] and McJunkin [26] give the nomograph of Fig. 88 as a method for determining handpump discharge. The method for using this is to rule a pencil line between the stroke length that applies (250mm or 10in in the example) and the expected pumping frequency (40 strokes per minute in the example). Then if another pencil line is ruled from where the first line crosses the "pivot line" through the appropriate cylinder diameter (76mm or 3in in the example), the discharge is given on the left (46 litre/min or 12 US gall/min in the example). No allowance is made for "slippage" or leakage of water which will result in the discharge being less than the swept volume, so the result of using this nomograph is the maximum flow that might be expected; it may therefore be more realistic to reduce the result obtained by 10-20%.

There have been many failures of handpumps in the field, especially in water supply projects where no particular individual readily takes responsiblity for the pump. In some countries communal handpumps developed a bad reputation as a result. Mention should be made of the major international UNDP/World Bank Global Handpump Project which seeks among other things to identify good handpumps for development use (mainly for village water supplies); this project has achieved a lot in this field and yielded a number of publications of relevance, notably [27].

Fig. 29 shows some of the key features of a typical handpump. Some of the weak points are the lever and fulcrum mechanism and the pump column itself; these can crack through metal fatigue or due to the use of poor quality castings. The pump in Fig. 29 is good in having a bracing strut to support the pump body, but it is bad in having the pivot bolt for the handle passing through the middle of the most highly stressed part of the pump lever. A better method of pivoting the hand lever is to have the pivot bearing passing through a lug below the lever arm to avoid weakening the arm at that point.

A further problem with hand lever pumps is wear and tear resulting from "hammering" in the drive train; this can be caused by worn pivots and bearings causing backlash, which will cause impacts if the operator lifts the handle too rapidly so it tries to overtake the piston and pump rod on the down stroke and then suddenly takes up the load again and also by users causing the hand lever to hit the end stops at the end of its travel. Also the need for the operator to constantly raise and lower a heavy lever (plus their arm(s)) wastes energy. Therefore rotary-drive pumps, in which the piston is driven by a crank from a rotating drive wheel are often easier to work (see Figs. 31 and 89). Here a flywheel smooths the fluctuations and thereby makes the pump easier to operate, especially for long periods, because the cyclic loading involved in accelerating discrete cylinder-volumes of water up the rising main will be absorbed by the flywheel's momentum and therefore not be felt by the operator. Rotary drive pumps of this kind have a further advantage as they cannot suffer from damage common with lever pumps from the operator hammering the end stops by moving the lever to far, or by the pump rod momentarily going "slack" due to the operator lifting the hand lever to rapidly.

The main disadvantage with rotary drive pumps is that they are generally relatively heavy and expensive due to their massive flywheel and

Fig. 89 Rotary drive hand pump (The Gambia)

Fig. 90 Rower pump (part sectioned) (after Klassen [28])

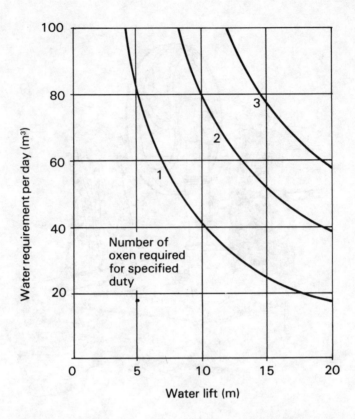

Fig. 91 The number of oxen required to provide a specified quantity of water at different lifts. The curves have been calculated by assuming than an ox can provide 350 watts of power for 5 hours per day, and the efficiency of the water-lifting device is 60%

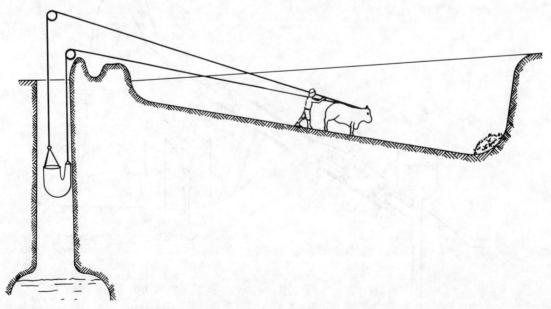

Fig. 92 Cross-section view of a Mohte (after Schioler [24])

crank mechanism, plus the supporting column that is needed. However, for pumping large quantities of water, as are required for irrigation, the improved ergonomics of the rotary drive is probably more advantageous than it is with water supply duties where no individual is likely to need to pump for more than a few minutes at any one time.

Ideally, any piston pump used for irrigation needs to use more than just the arm muscles. An attempt to produce such piston pump has been made in Bangladesh by the Mennonite Central Committee, with help from Caritas, according to Klassen; [28]. This pump is known as the "Rower Pump" because it is inclined at around 30° to the horizontal and operated with a rowing action (see Fig. 90). It is basically a simple and cheaply constructed pump, using 2" PVC pipe as the pump cylinder. The Rower Pump is claimed to pump 50% more water than a standard UNICEF No.6 lever hand pump used in MOSTI installations, (Manually Operated Shallow Tubewell for Irrigation), and the improvement is proportionately greater with higher lifts. However the main performance advantage of the rower pump is attributed by Klassen not so much to the action which is used (although this is claimed to make it easier to operate and might therefore allow longer periods of use per day) but because of a suction air chamber which smooths the flow into the pump (see Sections 3.5.4 and 3.7.2 for explanations of the use of air-chambers). Without the air chamber the performance falls to a level almost identical with that of the UNICEF pump, while if an air-chamber is fitted on the suction side of the UNICEF pump its performance is also enhanced to a similar level to that of the Rower Pump. Average performance quoted by Klassen for these two pumps, when pumping through heads in the range 5-6.5m (17-21ft), is as follows:

MCC Rower Pump		UNICEF No. 6 Pump		
without air chamber	with air chamber	without air chamber	with air chamber	
1.75	2.8	1.8	2.7	m³/h
7.7	12.3	8.0	11.7	US gall/min

This work is a good example of the kind of useful innovative developments that can be successfully pursued by NGOs and educational institutions in this important field, as suggested earlier,. It would be interesting to see the effect of adding air-chambers to the performance and ease of use of other types of handpump in use in the field.

4.2.4 Handpump Maintenance

The problem with industrially manufactured handpumps, as opposed to "self-build" devices like dhones or counterpoise lifts, is that they are dependent on spare parts which cannot easily be locally improvised. They also tend to need a certain amount of preventive maintenance if premature failure of components or impaired performance is to be avoided. However, they probably represent one of the only routes readily available to channel development funds into the widespread deployment of water-lifting equipment, since the alternative of using traditional hand-made devices is usually only practicable where there is an existing tradition make them and where the conditions for their use are right (eg. very low lifts from surface water).

Many water supply programmes using large numbers of handpumps have suffered serious difficulties with pump failures which have generally been attributed to poor maintenance. In some cases the failures have also been due to the use of poor pump designs which lack the capacity to survive intensive use. The kinds of problems experienced with hand pumps include:

a. poor quality of pump design and manufacture. This has partly resulted from manufacturers trimming the weight (and hence the cost) of components and generally degrading well tried designs in seeking to offer acceptably low bids in the absence of proper specifications to the procurement agencies;

b. iron and steel plain bearings and journals with poor fits and large clearances are provided; these properly should require very frequent lubrication which is impossible to provide, so rapid wear occurs;

c. the great variety of pumps in use leads to difficulties in finding the right spare parts;

d. very limited record keeping and feedback from the field makes it difficult to analyse the reasons for failures and to introduce remedial measures;

e. limited maintenance skills and equipment make it difficult for local people to undertake even basic overhaul operations, while lack of transport and poor communications make it difficult to summon help from a central source.

Attempts have been made to overcome some of these problems by introducing either a centralized system in which a maintenance team tours around repairing and maintaining a few dozen pumps in a district (this has often proved ineffective and expensive in practice). The other option being advocated is a "two tier" maintenance system, in which a central agency carries out the original installation, and provides a source of spare parts, training, transport, etc., but local people are trained to carry out routine repairs and maintenance.

4.3 ANIMAL POWER

The advantages of animal power over human power are twofold. First; draft animals are five to ten times more powerful than humans, so they can pump more water in a shorter time which tends to make the irrigation operation more efficient and productive. Second; by freeing the operator from having to work the water-lifting device, he can often manage the water distribution system more effectively. In effect, the use of an animal provides the equivalent power of several people, generally at a fraction of the cost.

Some 200 million draft animals are deployed in developing countries [29], and these have an aggregate power capacity of about 75,000,000kW and a capital value in the region of US $20 billion. The majority of these animals are used in southern and south east Asia; 80 million draft animals are in use in India alone. Any mechanization programme to replace these animals would obviously have to be extremely large; but there is more immediate scope for improving the efficiency with which animals are used.

Animal powered irrigation is almost exclusively practised using traditional water lifting techniques pre-dating the industrial era. Although some attempts to produce improved mechanisms for the utilisation of animal power have been made in certain areas during this century, there is little (if any) tendency to introduce animal powered water lifting anywhere where it has not been traditionally practised, although there seems good reason to believe it could usefully replace human labour for irrigation in many parts of the world where it is not already used. Instead, the trend has been either to make the quantum leap to full mechanisation using engines or mains electricity, or to make no attempt to improve on human power. An interesting exception is the "Water Buffalo Project" in Thailand where the use of buffalo powered persian wheels is being encouraged, [24].

The main disadvantage of animal power is that animals need to be fed for 365 days of the year, yet the irrigation season usually only extends for 100 or at most 200 days of the year. In areas where water is close to the surface, and hence irrigation by animal power is feasible, there are usually high human population densities and a shortage of land. Since draft animals consume considerable volumes of fodder, a significant proportion of the available land can be absorbed simply to support the draft animals. Therefore it probably would be difficult to justify the use of animals for irrigation pumping alone, but generally there are other economic applications for them, such as transport, tillage, and post-harvest duties like threshing or milling which allows them to be employed more fully than if they were used exclusively for irrigation. In India and other countries where animal powered water lifting is widely practised, it is normal for the same animals to be used for transport and for tilling the land. They are often fed with agricultural residues, or are allowed to graze on fields left fallow for a season as part of a crop rotation.

Animals also serve other purposes; they are a non-monetary form of collateral which is important in the village economy in many regions, they produce byproducts such as leather, meat and milk and of course in the Indian sub-continent in particular, their dung is widely used as a cooking fuel. So mechanization of irrigation pumps in lieu of draft animals is not necessarily a straightforward alternative.

4.3.1 Power capabilities of various species

Fig. 91 gives the approximate water lifting capability of different numbers of oxen, assuming 60% efficiency from animal to water lifted, as might be achieved by one of the better types of water lifting device.

The typical power capabilities of various commonly used draft animals are given in Table 13, based on [1] [30] [24] and other sources.

Draft animals obviously require rest just as humans do, so it is common practice to work them on about three hour shifts, with a rest in between. 10 or 12 hours per day in total may be worked when necessary.

Hood, [31], discussing typical draft horses as used in the USA at the end of the last century advocated the use of efficient mechanisms for coupling

Table 13   POWER AND DRAWBAR PULL OF VARIOUS ANIMALS

animal	weight (kg)	draft force (kg)	typical speed (m/s)	power (W)
heavy horse	700-1200	50-100	0.8 -1.2	500-1000
light horse	400-700	45-80	0.8 -1.4	400-800
mule	350-500	40-60	0.8 -1.0	300-600
donkey	150-300	20-40	0.6 -0.8	75-200
cow	400-600	50-60	0.6 -0.8	200-400
bullock/ox	500-900	60-80	0.5 -0.7	300-500
camel	500-1000	80-100	0.8 -1.2	400-700
buffalo	400-900	60-100	0.5 -1.0	600-1000

horses to water-lifts. He suggested that a team of two horses with an efficient water lifting device can typically lift 0.22 acre.feet per day through a head of 15ft, (270m³/day through 4.6m). He also cited various experiments completed at that time in Madras, India, (using bullocks), as an example for American farmers to emulate. In one of these, a device known as the Stoney Water Lift, was tested and shown to have an animal-to-water mechanical efficiency of about 80%. This device was, in effect, an engineered version of the traditional circular mohte with two balanced buckets, (see Fig. 93). Using a single Nellore bullock in a series of tests, it lifted between 2000 and 2500 Imp.Gall/hr through 22-23ft (9-11.25m³/h through 6.7-7.0m). It was pointed out that although a high instantaneous efficiency was achieved by this two bucket lift, in the tests it was found that the animals were only actually performing useful work for 60% of the duration of the tests, so the utilisation was not as high as may theoretically be feasible; clearly the management of an animal powered water lift has a major influence on the daily productivity.

An important point emphasised by Hood is that all draft animals work best when subjected to a steady load which matches their pulling capability, (although a horse, for example, can throw about one third of its weight into pulling for short periods - i.e. approaching three times the force it can sustain for a long time). Therefore, devices are needed which shield the animal from any cyclic loadings such as is experienced when a pump is driven by a crank, (see explanation of cyclic torque requirements of piston pump in Section 3.5).

4.3.2   Food Requirements

Birch and Rydzewski, [30], show that a cow in Bangladesh can be fed solely with forage and agricultural residues, (although the latter have a value equivalent to US $0.30 per day (1980) which is not a trivial sum) and that the same animal requires the residues from 0.77ha of double-cropped agricultural land. Because of the general shortage of land and residual fodder in Bangladesh, it is unusual to use animals for water lifting in that country.

The same authors carried out a similar calculation in relation to

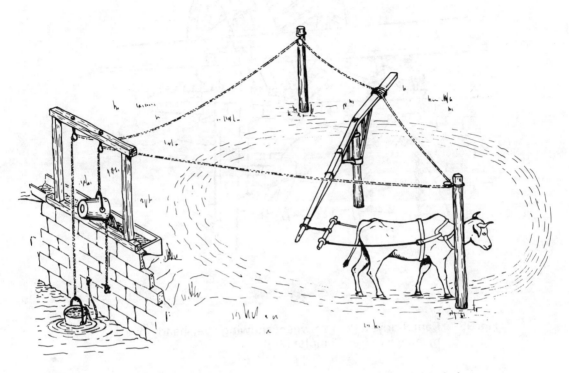

Fig. 93   Circular Mohte utilizing two buckets with flap-valves in bottom

Fig. 94   A bullock-driven Persian wheel of the conventional chain and bucket type which is widely used in many parts of the world, particularly the Near East and Southern Asia

Fig. 95 Camel-driven Persian wheel showing over-head drive mechanism (ref. Schioler [24])

Fig. 96 Animal-powered Chinese Liberation Pump
A hand-operated version is shown in Fig. 53 and a motorized one in Fig. 84

Egypt, where one hectare of land is needed to produce sufficient residue to support one animal, and the daily diet is valued at US $0.50.

4.3.3 Coupling animals to water lifting systems

The original method of using animals to lift water was some device such as the mohte, (Figs. 22 and 92). Here animals walk in a straight line down a slope away from the well or water source while hauling water up in a bag or container. Traditional mohtes used a leather bag to collect the water, but in recent years more durable materials such as rubber truck inner-tubes (or more rarely steel oil barrels) have been used.

The mohte is simple to implement; the only mechanical component is a pulley and the only structure is the frame to hold the pulley. However this key mechanical component is more complicated than might be expected due to the load on it, and demands considerable craftsmanship [24]. The instantaneous efficiency of the mohte is high while the animal(s) are pulling; a disadvantage is the need to reverse the animals back up the ramp to lower the bucket. Sometimes two teams of animals are used, so that one is led back to the top while one team descends; in order to do this, a man is needed at both ends of the ramp to harness and un-harness the animals at the end of each cycle. Using two sets of animals and, generally, two men and a boy, practically doubles the output (Molenaar, [1]), but even so inevitably no action occurs during the harnessing and un-harnessing process.

The downhill slope helps use the weight of the animal partially to balance the weight of the full water container; it also applies a reasonably constant load to the animal for most of the journey down the ramp, from when the container leaves the surface of the water to when it starts to be tipped.

An improved version of the mohte, sometimes used in parts of India and (a few) in Sri Lanka, is the "circular mohte"; Fig. 93. This involves attaching the animals to a sweep so that they can walk in a circle thereby allowing the animals to work continuously with less supervision. Because their weight can no longer partially balance the load, as they have to walk on level ground, two buckets are used so that the empty one descends while the full one comes up; this at least balances the weights of the buckets and means that the water being lifted is the only large out-of-balance force to be handled. The main problem with this device is that the load on the sweep is cyclic, as the pull on the sweep by the chain will not be felt by the animals when the chain acts parallel to the sweep and it will reach a maximum when the chain is at right angles to the sweep; therefore the animals will have a tiring sinusoidal load to cope with. Also, the various pulleys and supporting posts need to be robust and well anchored, as the forces are quite large.

The Persian wheel, (Figs. 23, 94 and 95) is a great improvement on the mohte, as its chain of buckets imposes an almost constant load on the drive shaft to the wheel. Persian wheels are usually driven by some form of right angle drive, such as in Figs. 94 and 95. The first is the most common, where the drive shaft from the secondary gear is buried and the animals walk over it; this has the advantage of keeping the Persian wheel as low as possible to minimise the head through which water is lifted. The second example is a traditional wooden Persian wheel mechanism where the animal passes under the horizontal shaft. The sweep of a Persian wheel carries an almost constant load and therefore the animal can establish a steady comfortable pace and needs little supervision.

In Egypt, the sakia (see Section 3.4.3 and Fig. 26) is commonly used for low head applications instead of the Persian wheel, and it is driven in a similar manner via a sweep. It has the advantage of applying a constant load to the animal, but is more efficient at very low heads.

The next development was for the sweep drive gear to be "industrialised" and manufactured in large numbers from iron or steel to include an engineered set of gears. Fig. 96 shows a mule harnessed to such a device in order to drive a chain and washer (or paternoster) pump which are widely used in China, where millions of liberation pumps, many of which were animal powered, were produced as an intermediate stage between human pumping and full mechanisation. The liberation pump includes a mechanism driven by a sweep that is elegantly simple and made from steel castings; (see Figs. 53 and 90). As detailed in Section 3.6.7, the liberation pump is capable of achieving an efficiency of approaching 70% in the animal powered form illustrated, and it is compact enough to fit into quite a narrow well.

Another not unusual concession to modernisation is the use of an old motor vehicle back-axle embedded in a concrete pillar as a means of obtaining a right angle drive from an animal sweep; an example of this is indicated in Fig. 97, where a donkey is shown linked to an Archimedian screw, (see also Section 3.6.3 and Fig. 48). Although in the example illustrated the matching problem is not always solved satisfactorily, according to Schioler, [24]. Again this is a compact and potentially efficient mechanism, as the Archimedean screw applies a completely steady load to the animal. It is clear that the same mechanism improvised from a car back axle could equally easily be coupled to a sakia, to a Persian wheel or to a chain and washer pump.

A new development in this field has been the appearance of a prototype animal powered, but industrially manufactured double-acting diaphragm pump from Denmark. This has the advantage of having the sweep direct coupled to the pump, which acts as a suction pump. Therefore it is only necessary to bury the pipes carrying the water rather than a drive shaft and this device can be located up to 80m away from the water source, which may in some cases be an advantage. This "Bunger" sweep pump is claimed to lift 100m^3/day using two animals.

Although a sweep driven device avoids the problem of reversing animals as with a mohte, it suffers from the disadvantage that by forcing an animal to walk in a circle, even though the load may be steady, the tractive effort or pull is reduced to 80% of that which is feasible when the same animal walks in a straight line; Hood; [31]. A mechanism which can apply a steady load to the animal, but in a straight line, would be better. The simplest device of this kind is the tread-wheel; in some respects the principle is analagous to a mohte, with the animal "walking on the spot" inside a large wheel, rather than up and down a ramp. This principle was taken further at the end of the nineteenth century in Europe and the USA through the use of "Paddle wheel" animal engines, in which a horse would be harnessed on an inclined endless belt which it would drive with its feet. The disadvantage of these animal carrying devices is that a mechanism is needed which not only has to transmit the maximum draw-bar pull of the animal, but additionally has to carry the full weight of the beast. Therefore a massive and robust construction is necessary which inevitably is expensive, and this probably is an example of where the search for maximum efficiency produces diminishing returns and therefore is counter-productive.

4.4 INTERNAL COMBUSTION ENGINES

The almost universal power source in the 5-500hp range, in areas having no mains electricity, is the internal combustion (i.c.) piston engine in either of its two main forms; the petrol (gasoline) fuelled spark ignition engine (s.i.) or the diesel fuelled compression ignition engine, (c.i.).

The main reasons for the widespread success of the i.c. piston engine are its high power/weight ratio, compact size and instant start-up capability, which led to its general adoption for powering motor vehicles in particular, and small isolated machinery and boats generally. Mass production, and decreasing fuel prices (until 1973) made small engines based on designs used for motor vehicles both cheap to buy and inexpensive to run. Since the 1973 and 1979 oil crises the trend of declining petroleum prices has generally reversed, although during the mid-1980s we have again a period of declining oil prices which may even last a few years. But even though crude oil prices have again declined (in US dollar terms), petroleum fuels frequently remain in short supply in many developing countries due to inability to finance sufficient oil imports. Almost everywhere the most serious supply shortages are most prevalent in the rural areas, in other words, precisely where farmers are in need of power for irrigation pumping. It is this actual or potential fuel supply problem which makes the alternatives to petroleum fuelled engines, reviewed later in this chapter, worth considering at all; the great versatility of the i.c. engine makes it exceedingly difficult to improve on it from any other point of view.

4.4.1 Different Types of I.C. Engines

This is not the book to describe the technicalities of small engines in detail; there are many standard textbooks on this subject and manufacturers and their agents will normally provide detailed information in the form of sales literature. However, when comparing the many available engines, it soon becomes apparent that there is a variety of different types of engine available even at one particular power output.

i. Diesel versus petrol/kerosene

The two main categories of engine to choose from are c.i. or diesel engines and s.i. or petrol (gasoline), kerosene or l.p.g. fuelled engines; (l.p.g. - liquified petroleum gas supplied in cylinders and usually primarily propane or butane). C.i. engines ignite their fuel by the heating effect when a charge of air is compressed suddenly enough; finely atomized fuel droplets are sprayed at very high pressure into the cylinder through the injector nozzle at the appropriate moment when the temperature is high enough to cause ignition. Spark ignition engines, on the other hand, work by mixing vapourized fuel with air, compressing the mixture and then igniting it at the correct moment by the electrical discharge of a spark plug set into the cylinder. Diesel engines therefore need to be heavier and more robust in construction to allow the high pressures needed to cause compression ignition to be sustained, and they also require a high pressure metering and injection pump to force the fuel in the right quantities at the right instant in time through the injector nozzle. The diesel injection pump and nozzles are built to a high level of precision and are therefore expensive components despite being mass-produced; they also depend on clean fuel and careful maintenance for reliable operation.

Therefore, petrol/gasoline and kerosene engines are inevitably lighter, more compact and usually cheaper than diesels. Although diesel engines are inherently more expensive to manufacture, they compensate for this by being more efficient, more reliable and more long-lasting (but more complicated to maintain in good running order). The main reasons for their better efficiency are firstly, the higher compression ratio allows a diesel to "breathe more deeply", to draw in more air per stroke in relation to the size of combustion space; secondly, fuel injection allows the diesel readily to run on a leaner fuel/air mixture than the equivalent s.i. engine. A spark ignition engine cannot be designed to run at such a high compression ratio, or the fuel/air mixture would ignite prematurely causing "knocking" or "pinking". Another less well known advantage of the diesel is that diesel fuel is 18% "richer" in energy than gasoline per litre (mainly due to its higher density); Table 14 indicates the calorific value of the three main petroleum fuels. Since fuel is generally bought by the litre (or some equivalent volume measure such as gallons), rather than by its weight or by its energy value, you can buy 18% more energy per litre of diesel than with petrol (gasoline).

Therefore the diesel is generally to be preferred as a power source, in terms of efficiency and reliable operation for long periods per day. For pumping applications, however, the choice of petrol or diesel relates largely to the scale of pumping required. Where a small, lightweight, portable system is needed which will only be used for one or two hours per day, and where simplicity of maintenance is important, and where "affordability" matters - i.e. the farmer has only the minimum capital to invest, then an i.c. gasoline or kerosene engine may be best and for that reason is frequently used.

It should be explained that the kerosene engine is similar to a petrol (gasoline) engine; indeed most kerosene engines need to be started and warmed up on petrol, because kerosene will not vapourize adequately in a cold engine. Many kerosene engines have a separate compartment in their fuel tank for a small supply of petrol and a tap to switch the fuel supply from petrol to kerosene once the engine is warm; it is also important to switch back to petrol a few moments before stopping the engine so that the carburettor float chamber is refilled with petrol ready for the next time the engine has to be started. Some farmers start kerosene engines simply by pouring petrol into the air intake, but this practice is not to be recommended as it can cause a fire. The advantage of a kerosene engine is that kerosene is normally available for agricultural purposes in an untaxed, subsidised or lightly taxed form and it also contains approximately 10% more energy per litre than petrol. The latter also usually carries a motor fuel tax in most countries as it is mainly used for private cars; therefore fuel costs for kerosene are generally much lower than for gasoline. Kerosene is also much less dangerous to store in quantity as it is much less easily ignited. The kerosene supply is also used for lighting and cooking fuel in many rural households and is therefore a more generally useful fuel.

Table 15 compares the general attributes of the three main i.c. engine options. It should be noted that diesels are sub-divided into two main categories; "low speed" and "high speed". The former run at speeds in the

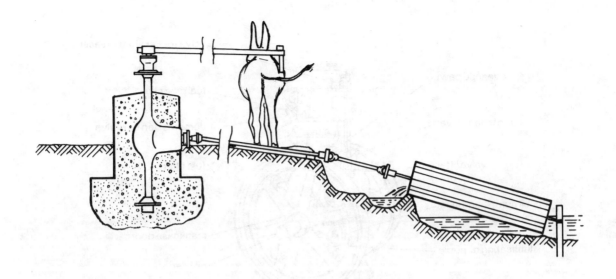

Fig. 97 The back axle from a car used as an animal-driven power transmission for an Archimedean screw pump (see also Fig. 48)

Table 14 COMPARISON OF DIFFERENT PETROLEUM-BASED ENGINE FUELS

Units	Petrol/ gasoline	Paraffin/ kerosene	Diesel oil/ gasoil
MJ/l	32	36	38
MJ/kg	44	45	46
kWh/l	9	10	11
kWh/kg	12	12	13
hp.h/US gall	45	51	54
hp.h/lb	4.1	4.2	4.5

Table 15 COMPARISON OF SMALL I.C. ENGINES

	Petrol/ gasoline	Paraffin/ kerosene	Diesel oil/gasoil High speed	Diesel oil/gasoil Low speed
Average fuel to shaft efficiency (%)	10-25	10-25	20-35	20-35
Weight per kW of rated power (kg)	3-10	4-12	10-40	20-80
Operational life (typical)	2000-4000h	2000-4000h	4000-8000h	8000-20000h
Running speed (rpm)	2500-3800	2500-3800	1200-2500	450-1200
Typical daily duty cycle (h)	0.5-4	3-6	2-10	6-24
Typical power ratings useful for small to medium irrigation (kW)	1-3	1-3	2-15	2-15

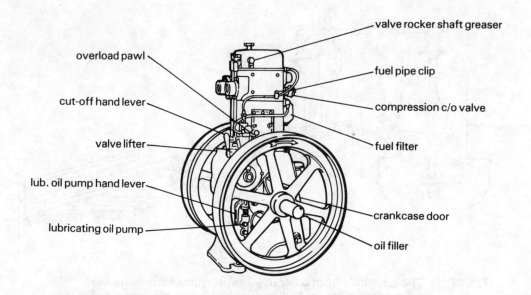

Fig. 98   Open flywheel low-speed single-cylinder diesel engine

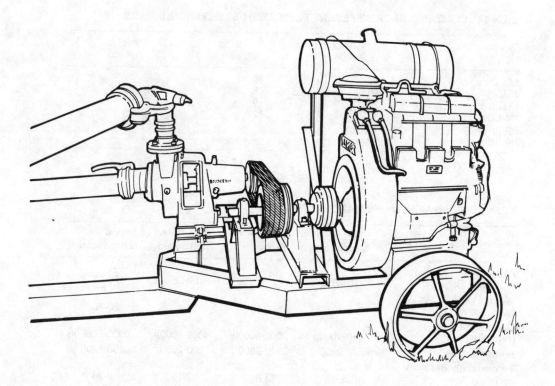

Fig. 99   Belt driven 3 cylinder Lister diesel engine coupled to a centrifugal pump via multiple 'V' belts

450-1200 rpm range and tend to be much heavier and more expensive in relation to their power rating than the latter which typically run at speeds in the 1200-2500 rpm range. The slow speed diesel tends to have a much longer operational life and to be better suited to continuous operation, or long duty cycles, but its initial purchase cost is much higher.

ii. Relationship between size, speed and durability

A general characteristic of all i.c. engines is that the smaller and lighter they are for a given power output, the lower will be their initial purchase price (which correlates to some extent with the weight) and the shorter will be their useful life. This is because a high power/weight ratio is normally achieved by running an engine at high speed; the faster an engine runs, the more air/fuel mixture it can consume and the greater will be the energy delivered. However, a faster machine will wear out quicker simply because its moving and rubbing components travel further in a given number of hours of use. There is therefore a tradeoff between heavy, expensive and slow engines on the one hand, and cheap and fast ones of the same power rating. Therefore, small, lightweight engines are recommended for such duties where portability and low first cost are important. In most cases, especially if the engine is part of a fixed installation and to be used for lengthy duty cycles, it will generally be worth investing in a suitably heavy and slower machine in the interests of achieving better reliability and a longer operational life. In general light s.i. engines are restricted to duties requiring less than 500 hours running time per season.

iii. Derating

If an engine is run continuously at its Rated Power, premature wear will occur. All engines therefore require to be derated from the manufacturer's rated power (which is the maximum power output the engine can achieve for short periods). Small engines are usually derated to about 70-80% of their rated power; eg. a 5kW rated engine will be necessary to produce a continuous 3.5-4.0kW.

The main reason for derating an engine is to prevent premature wear, but also the optimum efficiency for most engines is achieved at a speed corresponding to about 70-80% of its speed for maximum power. Therefore, derating an engine usually improves its specific fuel consumption (the fuel required per unit of output).

Further derating is necessary at high altitudes or at high ambient temperatures; recommendations to this effect are usually made by the manufacturer. Typically a further 10% derating is recommended for each 1,000m above sea level, plus 1% for each 5°C temperature rise above 16°C at the engine air intake. Therefore at 2,000m altitude and an ambient temperature of 26°C it would be necessary to derate an engine by say, 0.8 (generally) times 0.8 (for altitude) times 0.98 (for temperature) which totals .63 or 63% of rated power which would be the correct load to apply. Therefore a 2,000W load would require an engine nominally rated at 2,000/0.63 or 3.2kW (4.3bhp) under those conditions.

Excessive derating is to be avoided, as (particularly with diesels) running at a fraction of the design power tends to cause coking of the cylinder. Also, the engine efficiency will of course be much poorer than normal under such conditions.

iv.   Four-stroke or two-stroke

Both s.i. and c.i. engines can be designed to run so that ignition takes place either every other revolution, (four-stroke or four-cycle) or every revolution (two-stroke or two-cycle). The four-stroke s.i. engine tends to be more efficient as the "non-firing" revolution gives more time for inducing a fresh charge of fuel and also for effectively driving out the exhaust gases from the previous firing stroke (s.i. two-strokes tend to be less well scavenged of exhaust). Two-stroke diesels do not suffer an efficiency penalty in the same way, but are not generally available in the small size range of relevance for small scale irrigation. The two-stroke s.i. engine tends to be high revving and lightweight; it usually has fewer components than a four-stroke and therefore is cheaper to manufacture; typical applications are as moped engines. They are less suitable for irrigation pumping than four-strokes as they use more fuel and wear out more quickly. Most s.i. two-strokes use the downward movement of the piston into the crankcase to displace the air/fuel mixture into the cylinder; in which case it is not possible to lubricate the engine with a separate oil supply and the lubricant is mixed with the petrol, (two-stroke mix). This removes the need for oil changes, but is wasteful of lubricant, tends to cause a smoky exhaust, causes the need for more frequent "decokes" (de-carbonization of the cylinder head) and introduces a risk of damage caused by an inexperienced operator failing to mix sufficient lubricating oil with the fuel, or using the wrong type of oil. For these reasons, two-stroke s.i. engines are tending to be phased out and replaced by four strokes.

v.   Air or water cooling

About one third of the heat produced when the fuel is burnt has to be dissipated through the walls of the cylinder and through the cylinder head; the two methods generally used for removing this heat and preventing the cylinder overheating are either by surrounding the cylinder with a water jacket which has water circulated through it and a separate radiator, or by having many cooling fins on the cylinder (to increase its effective surface area), and blowing air over the fins with a fan driven off the engine. A few small, low-powered, and old fashioned low-speed stationery engines have a water jacket with an open top and keep cool simply by boiling the water, which needs to be topped up from time to time, but most modern liquid cooled engines have their coolant circulated by a pump just like car engines.

Each method of cooling has its pros and cons. Water cooled engines tend run slightly quieter and their engine temperature is more easily regulated through the use of a thermostat, than with air cooling. However, with water cooled engines, internal corrosion can occur and also water can leak out, evaporate or freeze. This last problem can be prevented by the use of anti-freeze (ethylene glycol) mixed with the cooling water in winter; most anti-freezes also contain corrosion inhibitor and are therefore useful to add to the cooling water even in climates where freezing is not likely to occur. Loss of coolant generally causes severe engine damage if the engine is allowed to continue running in that condition; various safety devices are available either to warn of overheating (caused by loss of coolant or for any other reason) and in some cases to automatically cut off the fuel supply and stop the engine. Air cooled engines obviously cannot lose their coolant, but it is important to ensure that their cooling fins do not get clogged with dust or dirt and that any cooling fan (when fitted) is clean and functioning correctly.

vi. **Cylinder arrangment**

The smallest engines usually have a single cylinder, mounted vertically above the crankshaft, as this is convenient for access to the main engine components and also allows an oil sump to be conveniently located where oil can drain down to it from the cylinder. A large flywheel is needed to smooth the output from a single cylinder engine as excessive vibration can cause problems with parts resonating and fatiguing and nuts and bolts working loose. Single cylinder low speed diesels need particularly heavy flywheels because they have large heavy pistons and connecting rods which run at low speeds; traditional designs are "open flywheel" (Fig. 98) while the more modern style of high-speed diesel engine usually has an internal enclosed flywheel (Fig. 99).

With larger sizes of engine it becomes feasible to have two or more cylinders. Twin cylinder engines have a smoother power output because the cylinders fire alternately and partially balance each other. Multi-cylinder engines therefore run more smoothly and quietly than ones of the same power with fewer cylinders, but the more cylinders there are the more components are involved, so obviously a multi-cylinder engine will be more expensive and more complicated to overhaul and maintain.

vii. **Special features and accessories**

Most small pumping systems have a hand crank starter or a pull chord (recoil) starter (the former is more common with small diesels and the latter with small petrol engines). Diesels often include a decompression valve to aid starting, in which a cylinder valve can be partially opened to release the pressure when the piston comes up on the compression stroke, allowing the engine to be hand wound up to a certain speed, when the decompression valve is suddenly closed and the momentum of the flywheel(s) carries the machine on sufficiently to fire the engine and start it off. Larger engines often have an auxilliary electrical system and a battery with an electric starter; (spark ignition engines generally need a battery to run the ignition, although very small s.i. engines use a magneto which generates and times the spark from the rotation of the engine). Engines fitted with electrical systems or engines coupled to generators can have various electrical controls to warn of or to prevent damage from overheating, loss of lubricant, etc. Some of these options, although sophisticated, are not expensive and are therefore a sound investment.

A vital, and often neglected accessory is the air filter, especially in dusty climates. Usually there is a choice of paper element filter (as used on most cars) where the paper element needs to be regularly replaced when clogged with dust or torn or, alternatively, an oil bath filter. The latter is slightly more expensive, but much more effective and practical in an agricultural context, since at a pinch even old engine oil can be used to refill it. A worn or malfunctioning air filter can greatly reduce the useful life of an engine, a fact which is often not fully understood by farmers judging from the number of engines that can be found running without any air filter at all.

viii.   Installation

The smallest engines are supplied mounted on skids or in a small frame and therefore need no installation other than coupling their pump to the water conveyance system. But larger machines, and many diesels, need to be properly installed, either on a concrete pad or on a suitable trolley or chassis. Most manufacturers will provide a detailed specification, when necessary, for the foundations of any engine driven pumping system; this should be accurately adhered to.

The engine often needs to be installed in a small lockable building for security. It is essential h,wever that any engine house is well ventilated and that the exhaust is properly discharged outside. This is not only to avoid the serious danger of poisoning the user with exhaust gases, but also to ensure the engine does not overheat. Similarly, engines with direct coupled centrifugal pumps sometimes need to be installed in a pit in order to lower them near enough to the water level to avoid an excessive suction lift. Considerable care is needed with such installations to ensure that neither exhaust nor oil fumes will fill the pit and poison anyone who enters it; carbon monoxide in i.c. engine exhaust emissions can, and frequently does, cause fatal accidents. Care is also needed to ensure that the water level will never rise to a level where it could submerge the engine.

4.4.2 Efficiency of Engine Powered Pumping Systems

This is a controversial subject where little reliable data on actual field performance exists in the literature. A few field tests have been carried out on "typical" irrigation pumping systems, and in some cases surprisingly poor efficiencies were achieved. For example, Jansen [32] reported on tests on three kerosene fuelled small pumping sets in the 2-3 bhp range in Sri Lanka; total system efficiencies in the range 0.75-3.5% were recorded although engine/pump efficiencies (without a pipeline) were in the range from 2.6-8.8%. Excessive pipeline friction losses obviously caused the very poor system performance. Unfortunately there is good reason to assume such losses are quite common.

Many of the reasons for poor performance can be corrected at little cost once it is recognized that a problem exists, but unfortunately, it is easy to run an inefficient pumping system without even realizing it, because the shortfall in output is simply made up by running the engine longer than would otherwise be necessary.

Fig. 100 indicates the principle components of a small engine pumping system and the range of efficiencies that can typically occur for each. Some explanation of these may give an insight into how such poor total system efficiency can sometimes occur, and by implication, what can be done to improve it.

Firstly, fuel spillage and pilferage could perhaps result in, anything from 0-10% of the fuel purchased being lost; i.e. 90-100% of fuel purchased may be usefully consumed as indicated in the figure. Spillage can occur not only when transferring fuel, but due to leaky storage or very frequently due to either a leaky fuel line on the engine or leaky joints especially on the high pressure lines of a diesel engine; well established fuel stains on the ground ought to give warning that something is wrong with fuel management. As the value of fuel increases, pilferage becomes increasingly a problem. One standard 200 litre drum of fuel is worth typically US $50-100; a small fortune in many developing countries.

Most basic thermodynamics textbooks claim that s.i. engines tend to be 25-30% efficient, while diesels are 30-40% efficient. Similarly, manufacturers' dynamometer tests, with optimally tuned engines, running on a test-bed, (often minus most of their accessories), tend to confirm this. However, such figures are optimistic in relation to field operations. The difference between theory and reality is greatest with the smallest sizes of engines, which are inherently less efficient than larger ones, and the text-book figures quoted above really only apply to well tuned engines of above 5-10kW power rating. The smallest two-stroke and four-stroke s.i. engines also tend to vary in quality, engine to engine, and can easily be as poor as only 10% efficient at around 1kW power rating. Small diesels (the smallest are generally about 1.5 to 2kW) will probably be better than 20% efficient as engines, but components like the injection pump, cooling fan and water circulating pump (all parasitic energy consumers) can reduce the fuel-to-useful-shaft-power efficiency to around 15% (or less) for the smallest engines. This large drop is because the parasitic accessories take proportionately more power from very small engines. Obviously engines are also only in new condition for a small part of their lives, and on average are worn and not well tuned, which also undermines their efficiency. Hence, depending on size, type and quality plus their age and how well they are maintained, engines may in reality be at best 35% efficient and at worst under 10% efficient.

The next source of loss is any mechanical transmission from the engine to the pump. In some cases the engine is direct coupled to the pump, in which case the transmission losses are negligible, but if there is any substantial change of speed, such as a speed-reducer and gearbox to drive a reciprocating borehole pump, then the transmission can be as low as 60% efficient, particularly with small systems of less than 5kW, where geaz box losses will be relatively large in relation to the power flow.

As discussed in the previous section, the most common type of pump used for irrigation with a small engine will be a centrifugal pump, either direct coupled or belt driven, usually working on suction; (eg. Figs. 72 and 99). If properly installed, so that the pump is operating close to its optimum head and speed, the pump efficiency can easily exceed 60% and possibly be as high as 80% with bigger pumps. However there is a lot of scope for failing to achieve these figures; bad impeller designs, worn impellers with much back-leakage, operating away from the design flow and head for a given speed will all have a detrimental effect and can easily singly, or in combination, pull the efficiency down to 25% or less. Given a reasonably well matched and well run system, pump efficiencies will therefore be in the 40-80% range, but they could easily be worse (but not better) than this range.

It is often not appreciated that the choice of delivery pipe can have a profound effect on system performance. Engines can deliver very high volumes at low heads, so pipe friction can grossly increase the total head across the pump, <u>particularly with long delivery lines at low heads</u>. When this happens, it is possible for the total head to be several times the static head, which multiplies the fuel requirement proportionately. Fig. 4 allows this to be quantified; eg. even a small portable petrol engine pump will typically deliver over 360 l/min or 6 l/s through 5m head (600W hydraulic output). The friction loss for each 100m of delivery pipe with this flow and head will be approximately as follows:

```
pipe nominal diameter:    2" (50mm)    2½"(65mm)    3" (80mm)
friction head (m/100m):   20m          5m           2m
```

from this, the pipe line efficiency for various lengths of pipe with the above diameters is as follows:

total pipe length	efficiencies in % for 5m static head and pipe diameters shown		
	2in	2½in	3in
10m	71	91	96
50m	33	67	83
100m	20	50	71

Quite clearly, when pumping at such low heads, it is easy to achieve total heads that are several times the static head, giving rise to pipeline efficiencies as poor as 20% when 100m of 2" pipe is used; (i.e. in that case the pumped head is _five times_ the static head so that five times as much fuel is needed compared with a 100% efficient pipeline). The situation gets proportionately less serious at higher static heads, because it is the ratio of pipe friction head to static head that matters; for example, at 20m static head the above example would give the same friction head of 20m which although unacceptably high would at least imply 50% rather than 20% pipeline efficiency. It is a common mistake to use pipework which is too small in diameter as a supposed "economy" when larger pipe can often pay for itself in saved fuel within months rather than years. Also, some pumps which have a 2in discharge orifice may actually need a 3in pipeline if some distance is involved, yet uninformed users will usually use a pipe diameter to match the pump orifice size and thereby create a major source of inefficiency. Incidentally, inch pipe sizes are quoted here simply because they are in fact more commonly used, even in countries where every thing else is dimensioned in metric units.

The performance curve in Fig. 101 indicates how centrifugal suction pumps can also suffer reduced performance as the suction head increases, mainly due to cavitation, particularly when the suction head is a large fraction of the total head. The figure shows how at 10m total head, 6m suction head causes a 20% drop in output compared with 3m suction head. This is without any reduction in power demand, so the former system is 20% less efficient than the latter, simply due to suction losses. There is therefore a potentially large fuel cost penalty in applying excessive suction lifts (apart from the usual priming problems that can occur).

Pipe losses, whether due to friction or suction can therefore cause increased head and reduced flow and will typically have an efficiency in the 30-95% range, as indicated in Fig. 100.

The efficiency factors discussed so far apply to the hardware, while it is being run and water is being usefully applied to the field. Inevitably the system needs to run when water is not being usefully applied. For example, when starting up, any engine will often be run for a few minutes

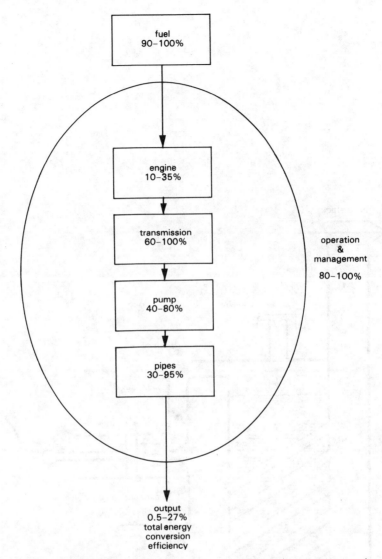

Fig. 100 Principal components of a small engine pumping system and their efficiencies

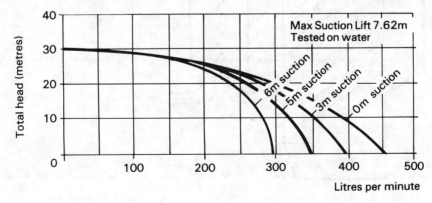

Fig. 101 Effect of increasing the suction head on the output for a typical engine pump set

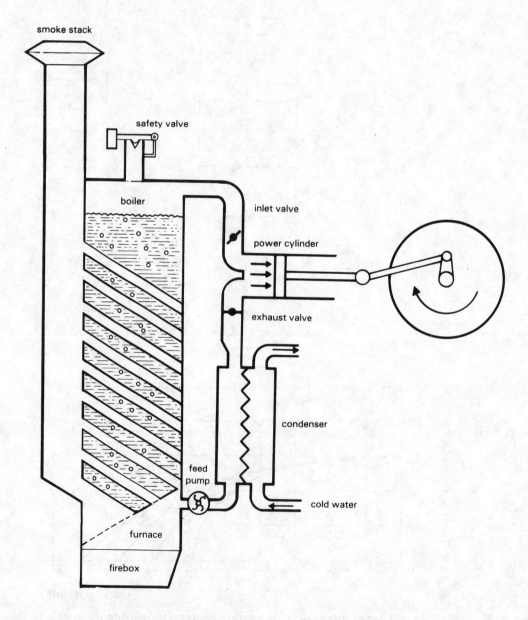

Fig. 102 Schematic arrangement of a condensing steam engine

before the farmer can arrange the discharge to reach the correct part of the field, and water will be wasted. Similarly when rearranging the distribution system to deliver water to another part of the field, some wastage may occur. There is therefore a factor relating to the type of water distribution system and to the management skill of the farmer at applying the pumped water for as large a fraction of the time it is being pumped (or the engine is running) as possible. Even moderately bad management could cause 20% loss compared with ideal usage, and really bad management can be much worse, so this efficiency is taken as ranging from 80-100% for the purpose of Fig. 100.

When all the worst efficiencies suggested in Fig. 100 are compounded, they yield a theoretical worst total efficiency of 0.5% (which Jansen [32] and others have confirmed) while the best factors in Fig. 100 compounded together give 27%. The "best" figure is only even theoretically feasible however for a larger diesel (over 5kW), driving a pump at 10-20m (or higher) head, which is rather higher than usual for most irrigation applications. Most smaller engine pumped irrigation systems therefore in practice probably achieve 5 to 15% total efficiency, with larger diesels operating at higher heads being towards the top end of this range and small kerosene or petrol engines at low heads being at (or below) the bottom. The operator should probably be satisfied if a small diesel system achieves 10-15% efficiency and a small petrol or kerosene fuelled system achieves 5-10%. It may be very worthwhile for any reader to investigate the actual efficiency of any engine pumping system they may be responsible for, (by comparing fuel consumption against hydraulic energy output) so that if it is below par steps may be taken to find the causes and to correct them.

4.5 EXTERNAL COMBUSTION ENGINES

The difference between internal and external combustion engines, as their names suggest, is that the former burn their fuel within the power cylinder, but the latter use their fuel to heat a gas or a vapour through the walls of an external chamber, and the heated gas or vapour is then transferred to the power cylinder. External combustion engines therefore require a heat exchanger, or boiler to take in heat, and as their fuels are burnt externally under steady conditions, they can in principle use any fuel that can burn, including agricultural residues or waste materials

There are two main families of external combustion engines; steam engines which rely on expanding steam (or occasionally some other vapour) to drive a mechanism; or Stirling engines which use hot air (or some other hot gas). The use of both technologies reached their zeniths around 1900 and have declined almost to extinction since. However a brief description is worthwhile, since:

i. they were successfully and widely used in the past for pumping water;

ii. they both have the merit of being well suited to the use of low cost fuels such as coal, peat and biomass;

iii. attempts to update and revive them are taking place.

and therefore they may re-appear as viable options in the longer term future.

The primary disadvantage of e.c. engines is that a large area of heat exchanger is necessary to transmit heat into the working cylinder(s) and also to reject heat at the end of the cycle. As a result, e.c. engines are generally bulky and expensive to construct compared with i.c. engines. Also, since they are no longer generally manufactured they do not enjoy the economies of mass-production available to i.c. engines. They also will not start so quickly or conveniently as an i.c. engine; because it takes time to light the fire and heat the machine to its working temperature.

Due to their relatively poor power/weight ratio and also the worse energy/weight ratio of solid fuels, the kinds of applications where steam or Stirling engines are most likely to be acceptable are for static applications such as as irrigation water pumping in areas where petroleum fuels are not readily available but low cost solid fuels are. On the positive side, e.c. engines have the advantage of having the potential to be much longer-lasting than i.c. engines (100 year old steam railway locomotives are relatively easy to keep in working order, but it is rare for i.c. engines to be used more than 20 years or so. E.c. engines are also significantly quieter and free of vibrations than i.c. engines. The level of skill needed for maintenance may also be lower, although the amount of time spent will be higher, particularly due to the need for cleaning out the furnace.

Modern engineering techniques promise that any future steam or Stirling engines could benefit from features not available over 60 years ago when they were last in general use. Products incorporating these new developments are not yet on the market, but R&D is in hand in various countries on a limited scale; however it will probably be some years before a new generation of multi-fuel Stirling or steam powered pumps become generally available.

4.5.1 Steam Engines

Only a limited number of small steam engines are available commercially at present; most are for general use or for powering small pleasure boats. A serious attempt to develop a 2kW steam engine for use in remote areas was made by the engine designers, Ricardos, in the UK during the 1950s (see Fig. 157). That development was possibly premature and failed, but there is currently a revival of interest in developing power sources that can run on biomass-based fuels (as discussed more fully in Section 4.10). However, small steam engines have always suffered from their need to meet quite stringent safety requirements to avoid accidents due to boiler explosions, and most countries have regulations requiring the certification of steam engine boilers, which is a serious, but necessary, inhibiting factor.

The principle of the steam engine is illustrated in Fig. 102. Fuel is burnt in a furnace and the hot gases usually pass through tubes surrounded by water (fire tube boilers). Steam is generated under pressure; typically 5 to 10 atmospheres (or 5-10bar). A safety valve is provided to release steam when the pressure becomes too high so as to avoid the risk of an explosion. High pressure steam is admitted to a power cylinder through a valve, where it expands against a moving piston to do work while its pressure drops. The inlet valve closes at a certain point, but the steam usually continues expanding until it is close to atmospheric pressure, when the exhaust valve opens to allow the piston to push the cooled and expanded steam out to make way for a new intake of high pressure steam. The valves are linked to the drive mechanism so as to open or close automatically at the correct moment. The period of opening of the inlet valve can be adjusted by the operator to vary the speed and power of the engine.

In the simplest types of engine the steam is exhausted to the atmosphere. This however is wasteful of energy, because by cooling and condensing the exhausted steam the pressure can be reduced to a semi-vacuum and this allows more energy to be extracted from a given throughput of steam and thereby significantly improves the efficiency. When a condenser is not used, such as with steam railway locomotives, the jet of exhaust steam is utilised to create a good draught for the furnace by drawing the hot gases up the necessarily short smoke stack. Condensing steam engines, on the other hand, either need a high stack to create a draught by natural convection, or they need fans or blowers.

Steam pumps can easily include a condenser, since the pumped water can serve to cool the condenser. According to Mead [13], (and others) the typical gain in overall efficiency from using a condenser can exceed 30% extra output per unit of fuel used. Condensed steam collects as water at the bottom of the condenser and is then pumped at sufficient pressure to inject it back into the boiler by a small water feed pump, which is normally driven off the engine. A further important advantage of a condensing steam engine is that recirculating the same water reduces the problems of scaling and corrosion that commonly occur when a continuous throughput of fresh water is used. A clean and mineral-free water supply is normally necessary for non-condensing steam engines to prolong the life of the boiler.

The most basic steam engine is about 5% efficient (steam energy to mechanical shaft energy - the furnace and boiler efficiency of probably between 30 and 60% needs to be compounded with this to give an overall efficiency as a prime-mover in the 1.5 to 3% range). More sophisticated engines are around 10% efficient, while the very best reach 15%. When the boiler and furnace efficiencies (30-60%) plus the pump (40-80%) and pipework (40-90%) are compounded, we obtain system efficiencies for steam piston engine powered pumps in the 0.5 to 4.5% range, which is worse, but not a lot worse than for small s.i. internal combustion engines pumping systems, but allows the use of non-petroleum fuels and offers greater durability.

4.5.2 Stirling Engines

This type of engine was originally developed by the Rev. Robert Stirling in 1816. Tens of thousands of small Stirling engines were used in the late nineteenth and early twentieth century, mainly in the USA but also in Europe. They were applied to all manner of small scale power purposes, including water pumping. In North America they particularly saw service on the "new frontier"; which at that time suffered all the problems of a developing country in terms of lack of energy resources, etc.

Rural electrification and the rise of the small petrol engine during and after the 1920s overtook the Stirling engine, but their inherent multi-fuel capability, robustness and durability make them an attractive concept for re-development for use in remote areas in the future and certain projects are being initiated to this end. Various types of direct-action Stirling-piston water pumps have been developed since the 1970s by Beale and

Sunpower Inc. in the USA, and some limited development of new engines, for example by IT Power in the UK with finance from GTZ of West Germany is continuing.

Stirling engines use pressure changes caused by alternately heating and cooling an enclosed mass of air (or other gas). The Stirling engine has the potential to be more efficient than the steam engine, and also it avoids the boiler explosion and scaling hazards of steam engines. An important attribute is that the Stirling engine is almost unique as a heat engine in that it can be made to work quite well at fractional horsepower sizes where both i.c. engines and steam engines are relatively inefficient. This of course makes it of potential interest for small scale irrigation, although at present it is not a commercially available option.

To explain the Stirling cycle rigorously is a complex task. But in simple terms, a displacer is used to move the enclosed supply of air from a hot chamber to a cold chamber via a regenerator. When most of the air is in the hot end of the enclosed system, the internal pressure will be high and the gas is allowed to expand against a power piston, and conversely, when the displacer moves the air to the cool end, the pressure drops and the power piston returns. The gas moves from the hot end to the cold end through a regenerator which has a high thermal capacity combined with a lot of surface area, so that the hot air being drawn from the power cylinder cools progressively on its way through the regenerator, giving up its heat in the process; then when cool air travels back to the power cylinder ready for the next power stroke the heat is returned from the regenerator matrix to preheat the air prior to reaching the power cylinder. The regenerator is vital to achieving good efficiency from a Stirling engine. It often consists of a mass of metal gauze through which air can readily pass, [33], [34].

Some insight into the mechanics of a small Stirling engine can be gained from Fig. 103, which shows a 1900 vintage Rider-Ericsson engine. The displacer cylinder projects at its lower end into a small furnace. When the displacer descends it pushes all the air through the regenerator into the water cooled volume near the power cylinder and the pressure in the system drops, then as the displacer rises and pulls air back into the hot space, the pressure rises and is used to push the power piston upwards on the working stroke. The displacer is driven off the drive shaft and runs 90° out of phase with the power piston. An idea of the potential value of engines such as this can be gained from records of their performance; for example, the half horsepower Rider-Ericsson engine could raise 2.7m³/hr of water through 20m; it ran at about 140 rpm (only) and consumed about 2kg of coke fuel per hour. All that was needed to keep it going was for the fire to be occasionally stoked, rather like a domestic stove, and for a drop of oil to be dispensed onto the plain bearings every hour or so.

4.6   ELECTRICAL POWER

If a connection is available to a reliable mains electricity supply, nothing else is either as convenient or more cost-effective for powering an irrigation pump. Unfortunately, the majority of farmers in developing countries do not have mains electricity close at hand, and even those that do often find that the supply is unreliable. Electricity supply problems tend to be particularly prevalent during the irrigation season, because irrigation pumping tends to be practised simultaneously by all farmers in a particular district and can therefore easily overload an inadequate rural network and cause "brownouts" (voltage reductions) or even "blackouts" (complete power cuts). Therefore there is a major inhibition for many electricity utilities in encouraging any further use of electricity for irrigation pumping in developing countries where the electrical supply network is already under strain.

The real cost of extending the grid is very high, typically in the order of $5,000-10,000 per kilometer of spur. Although connections in many countries have in the past been subsidised, whatever the pricing policy of the utility, someone has to pay for it and the tendency today is to withdraw subsidies. Therefore, although an electric motor considered in isolation is an extremely inexpensive and convenient prime-mover, it is only useful when conbnnected to a lot of capital-intensive infrastructure which needs to carry a substantial electrical load in order to be self-financing from revenue.

A further problem for developing countries in considering the mains electricity option is the high foreign exchange component in the investment; this is typically from 50-80%, according to Fluitman, [35] (quoting a World Bank source). Electricity generation in rural areas of developing countries tends to be by petroleum-fuelled plant (usually diesel generators) so this also is a burden on the economy. In fact a large fraction of many developing countries' oil imports goes to electrical power generation. The attractiveness of rural electrification as an investment for development is therefore being questioned much more now then it used to be; (eg. [35]). However it is not proposed here to deal with policy implications or macro-economic effects of the widespread use of electricity for irrigation pumping, other then to point out that it cannot be seen as a universally applicable solution to the world's irrigation pumping needs, because most countries will not be able to afford to extend a grid to all their rural areas in the forseeable future. Even where such an option can be afforded, it is still necessary to question whether it is the most cost-effective solution for irrigation pumping bearing in mind the high infrastructural costs.

4.6.1 Sources and Types of Electricity

Batteries produce a steady flow of electricity known as "direct current" or DC. Photovoltaic (solar) cells also produce DC. Electrical generators to produce DC are sometimes known as "dynamos"; they require commutators consisting of rotating brass segments with fixed carbon brushes. Alternators are almost universally used today for the generation of electricity from shaft power. Alternators are simpler and less expensive than DC generators, but they produce a voltage which reverses completely several times per revolution. This type of electrical output, which is almost universally used for mains supplies, is known as "alternating current" or AC.

AC mains voltage normally fluctuates from full positive to full negative and back 50 times per second (50Hz or 50 cycles/sec) or in some cases at 60Hz. The current fluctuates similarly. Sometimes the current and voltage can be "out of step", i.e. their peaks do not coincide. This discrepency (or phase difference) is quantified by the "power factor"; the output of an AC system is the product of the amps, volts and the power factor. When the amps and volts are in perfect phase with each other, the power factor is numerically 1. When the power factor is less than one (it frequently is 0.9 and sometimes less) then the power available is reduced proportionately for a given system rating. The rating of AC equipment is therefore generally given not in watts or kilowatts (kW), but in volt-amps or kilovolt-amps (kVA). The actual power in kW will therefore be the kVA rating multiplied by the power factor.

Another important principle to be aware of is that it is considerably more economic to transmit electricity any distance at high voltages rather than low. A smaller cross-section of conductor is needed for a given transmission efficiency. This is analagous to water transmission, where higher pressures and smaller flow rates allow smaller pipes to be used for equal hydraulic power. However, electricity is potentially lethal at AC voltages much above 240V and at DC voltages much above 100V (it can of course kill at considerably lower voltages depending on the circumstances and state of health of the victim) and insulation becomes more difficult. Therefore, for safety reasons, 240V AC or about 110V DC are usually the maximum voltages used at the end-users' supplies and for electrical appliances.

The reason AC is generally used for mains applications rather than DC are that it has a number of important advantages:

a. AC generators and motors are much simpler, less expensive and less troublesome, since they do not require commutators;

b. AC voltages can be changed efficiently and with a high degree of reliability, using transformers, but it is a technically much more difficult problem to change DC voltages; therefore AC can easily be transmitted efficiently at high voltages and then transformed to low, safer voltages close to the point of use;

c. as a result of the advantages of AC, it has become the internationally used standard for mains supplies and virtually all mass-produced electrical applicances are designed for AC use.

It is sometimes necessary to convert AC to DC or vice-versa, for example to charge batteries (which are DC) from the AC mains or to run an AC appliance designed for the mains from a DC source such as a battery or a solar photovoltaic array. AC can quite readily be converted to DC by using a rectifier; these (like transformers) are solid-state devices which require no maintenance and are relatively efficient. A battery charger usually consists of a combination of a transformer (to step mains voltage down to battery voltage) and a rectifier to convert the low voltage AC to DC. Converting DC to AC is more difficult; traditionally an inefficient electro-mechanical device called a rotary converter was used; this is a DC motor direct coupled to an AC alternator. The modern alternative is an electronic, solid state device called an inverter. Inverters are relatively inexpensive for low power applications (such as powering small fluorescent lights from low voltage batteries), but they become expensive for such higher powered applications such as electric motors for pumping. The quality and price of inverters also varies a lot; if a good quality AC output is essential (and high efficiency of conversion) a more complicated and expensive device is needed. Cheap inverters often produce a crude AC output and are relatively inefficient; they can also seriously interfere with radio and TV reception in the vicinity.

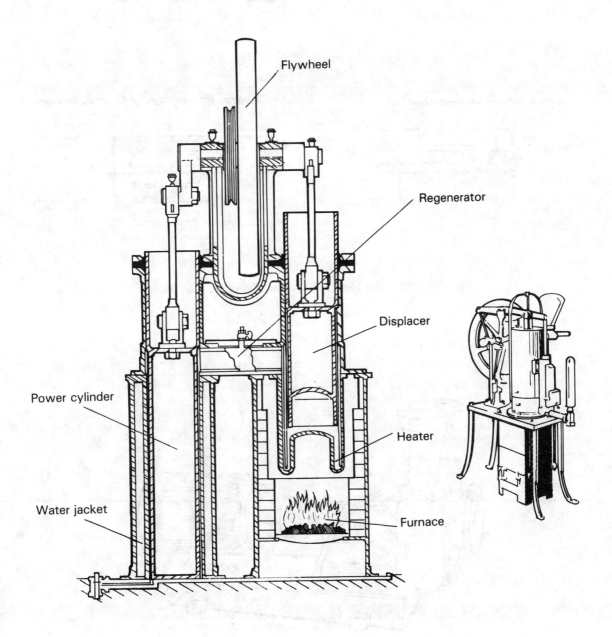

Fig. 103 Rider-Ericsson hot air pumping engine (Stirling cycle) circa 1900

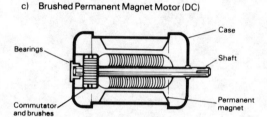
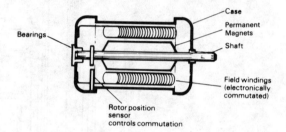

Fig. 104 The four main types of electric motors

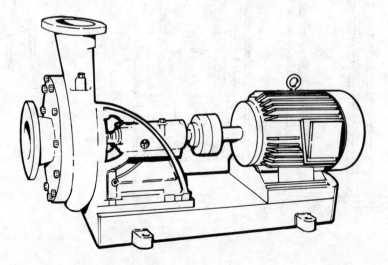

Fig. 105 Direct coupled electric motor and centrifugal pump

4.6.2 AC Mains Power

Mains electricity is generally supplied as alternating current (AC) either at 220 to 240V and 50Hz frequency or at 110V and 60Hz frequency for low power connections, (including domestic ones) of up to about 10kW. The 220-240V 50Hz standard is normal in Europe while the 110V 60Hz standard is in use in the USA; either might be used in other parts of the world, although 220-240V is more common, especially in Asia and Africa.

When AC is supplied through two wires, it is known as single-phase. The two wires are not "positive" and "negative" but are "live" and "neutral"; there should always also be a third wire included for safety - the "earth" or "ground". The latter is normally connected to the casing of any appliance or motor so that if any internal fault causes the casing to come into contact with the live supply, the leakage current will flow to earth (ground) and trip out the system or blow a fuse. Therefore if an electric pump keeps tripping or blowing fuses it is as well to have it checked to see if there is a short-circuit.

Mains power is normally generated as "three-phase", in which the alternator transmits three "single phase" AC outputs down three wires. Each phase is shifted by one third of a revolution of the alternator, so the voltage peaks in the three conductors do not coincide, but are evenly spaced out. The three phases, if equally balanced, will cancel each other out if fed through three equal loads, but in practice they are not usually perfectly balanced so there is normally a fourth return conductor called the neutral. A single phase AC supply is simply a connection to one of the three "lives" of a three-phase source with a return to its neutral. For this reason it is important in many cases not to confuse the live and neutral; also it is the live which should be protected by fuses or contact breakers.

At higher power levels, usually above 5kW, and always above about 25kW, it is normal to use three-phase AC. This is supplied mostly at 415V line to line (Europe) or 190 or 440V (USA).

4.6.3 Electric Motors

An electric motor seems almost the ideal prime mover for a water pump. Power is supplied "at the flick of a switch", and water is produced at a constant rate until the motor is turned off. Electric motors have relatively long service lives and generally need little or no servicing.

The cheapest and simplest type of electric motor is the squirrel cage induction motor which is almost universally used for mains electric power applications; see Fig. 104 (a) and Fig. 105. Here there are no electrical connections to the rotating "squirrel cage", so there are no brushes or slip rings to wear or need adjustment. Motors of this kind are available in either three-phase or single-phase versions. They run at a fixed speed depending on the frequency of the power supply and the number of poles in their stator windings. The most general type (which is usually the cheapest) runs at a nominal 1,500rpm at 50Hz (1,800 at 60Hz), but other speeds are available. It is normal to direct-couple a motor to a centrifugal pump where possible (eg. Fig 105). Non-standard speed motors may be used where this does not suit the pump, or alternatively a belt speed reduction arrangement may be used, such as in Fig 106.

A problem with induction motors is that they normally need over three times as much current to start as they do once running at rated speed and power. This means that the peak current that can be supplied must be significantly higher than that needed for operation, which often causes not just technical but also financial problems, as some electricity tariffs are determined by the maximum current rating on a circuit. Recently electronic starting devices have become available which limit the starting current while the motor runs up to speed and which in some cases also improve the overall efficiency of an electric motor.

Induction motors are typically 75% efficient for a 300W (0.5 hp) size and may be around 85% efficient at 10kW size (subject to having a unity power factor). They are not generally made in sizes significantly smaller than 100-200W.

For very small scale applications, the so-called "universal motor" is most commonly used. The universal motor (Fig. 104 (b)) is the "classic" electric motor with a brushed commutator and wound armature. Fixed field coils produce the magnetic flux to run the motor. Motors of this kind can use either an AC or a DC supply and they are typically used for very small-scale power applications (such as in power tools, and domestic appliances, for example). They are more efficient than would be possible with a very small induction motor and their starting current is smaller in relation to their running current. They suffer however from needing periodic replacement of brushes when used intensively, as for pumping duties.

There are small-scale electrical power applications independent of a mains supply, which use a DC source such as a photovoltaic array, or batteries charged from a wind-generator. In these applications a permanent magnet DC motor is the most efficient option (Fig. 104 (c)). In these, permanent magnets replace the field coils; this offers higher efficiency, particularly at part-load, when field windings would absorb a significant proportion of the power being drawn. Permanent magnet DC motors can be 75-85% efficient even at such low power ratings as 100-200W, needed for the smallest solar pumping systems. Most permanent magnet motors have brushed/commutated armatures exactly like a universal motor, which in the pumping context is a major drawback particularly for submersible sealed in motors. However brushless permanent magnet motors have recently become available (Fig 104 (d)). Here the magnets are fixed to the rotor and the stator windings are fed a commutated AC current at variable frequency to suit the speed of rotation; this is done by sending a signal from a rotor position-sensor which measures the speed and position of the shaft and controls electronic circuitry which performs the commutation function on a DC supply. Motors of this kind are mechanically on a par with an induction motor, and can be sealed for life in a submersible pump if required, but they are still produced in limited numbers and involve a sophisticated electronic commutator which makes them relatively expensive at the time of writing. With the increasing use of solar pumps they are likely to become more widely used and their price may fall.

Submersible pump motors, whether AC induction motors or DC brushless permanent magnet motors, are commonly filled with (clean and corrosion inhibited) water as this equalises the pressure on the seals and makes it easier to prevent ingress of well water than if the motor contained only air at atmospheric pressure. Filling motors with water is obviously only possible with brushless motors, or short-circuits would occur. Another advantage of water filled motors is that they are better-protected from overheating.

4.6.4 Electrical Safety

AC electrical voltages over about 110V and DC over 80V are potentially lethal, especially if the contact is enhanced through the presence of water. Therefore, electricity and water need to be combined with caution, and anyone using electricity for irrigation pumping should ensure that all necessary protection equipment is provided; i.e. effective trips or fuses, plus suitable armoured cables, earthed and splashproof enclosures, etc. Also all major components, the motor, pump and supporting structure should be properly earthed (or grounded) with all earth connections electrically bonded together. It is vital that electrical installations should either be completed by trained electricians or if the farmer carries it out, he should have it inspected and checked by a properly qualified person before ever attempting to use it; (in some countries this is in any case a legal requirement). It is also prudent to have some prior knowledge what action to take for treating electric shock; most electrical utilities can provide posters or notices giving details of precautions with recommendations on treatment should such an unfortunate event occur.

4.7 WIND POWER

4.7.1 Background and State-of-the-Art

i. Background

The wind has been used for pumping water for many centuries; it was in fact the primary method used for dewatering large areas of the Netherlands from the 13th century onwards; [36]. Smaller windpumps, generally made from wood, for use to dewater polders, (in Holland) and for pumping sea water in salt workings, (France, Spain and Portugal), were also widely used in Europe and are still used in places like Cape Verde; Fig. 107.

However the main type of windpump that has been used is the so-called American farm windpump; (Fig. 108). This normally has a steel, multibladed, fan-like rotor, which drives a reciprocating pump linkage usually via reduction gearing (Fig. 109) that connects directly with a piston pump located in a borehole directly below. The American farm windpump evolved during the period between 1860 and 1900 when many millions of cattle were being introduced on the North American Great Plains. Limited surface water created a vast demand for water lifting machinery, so windpumps rapidly became the main general purpose power source for this purpose. The US agricultural industry spawned a multitude of windpump manufacturers and there were serious R&D programmes, some sponsored by the US government, [37], to evolve better windpumps for irrigation as well as for water supply duties.

Other "new frontiers" such as Australia and Argentina took up the farm windpump, and to this day an estimated one million steel farm windpumps are in regular use [38], the largest numbers being in Australia and Argentina; [39], [40]. It should be noted that the so-called American Farm Windpump is rarely used today for irrigation; most are used for the purpose they were originally developed for, namely watering livestock and, to a lesser extent, for farm or community water supplies. They tend therefore to be applied at quite high heads by irrigation standards; typically in the 10 to 100m range on boreholes. Large windpumps are even in regular use on boreholes of over 200m depth.

Wind pumps have also been used in SE Asia and China for longer than in Europe, mainly for irrigation or for pumping sea water into drying pans for sea salt production. The Chinese sail windpump (Fig. 110) was first used over a thousand years ago and tens if not hundreds of thousands, are still in use in Hubei, Henan and North Jiangsu provinces [41]. The traditional Chinese designs are constructed from wire-braced bamboo poles carrying fabric sails; usually either a paddle pump or a dragon-spine (ladder pump) is used, typically at pumping heads of less than 1m. Many Chinese windmills rely on the wind generally blowing in the same direction, because their rotors are of fixed orientation. Many hundreds of a similar design of windpump to the Chinese ones are also used on saltpans in Thailand, (Fig. 111).

Some 50,000 windpumps were used around the Mediterranean Sea 40 years ago for irrigation purposes, [42]. These were improvised direct-drive variations of the metal American farm windpump, but often using triangular cloth sails rather than metal blades. These sail windmills have a type of rotor which has been used for many centuries in the Mediterranean region, but today is often known as "Cretan Windmills" (see Fig. 112). During the last 30 years or so, increased prosperity combined with cheaper engines and fuels has generally led farmers in this region to abandon windmills and use small engines (or mains electricity where available). However Crete is well known as a country where until recently about 6,000 windpumps were still in use [91], mostly with the cloth sailed rig. The numbers of windpumps in use in Crete are rapidly declining and by 1986 were believed to be barely one thousand.

Another branch of wind energy technology began to develop in the late 1920s and early 1930s, namely, the wind-generator or aero-generator. Many thousands of small wind generators, such as the Australian Dunlite (Fig. 113), were brought into use for charging batteries which could be used for lighting, and especially for radio communication, in remote rural areas. Such machines can also provide an alternative to a photovoltaic array for irrigation pumping in suitably windy areas, although they have not so far been applied for this purpose in any numbers.

Large wind turbines for electricity generation have been (and are being) constructed, the largest being a 5MW (5,000kW) machine under development in West Germany. However, more modest, but still quite large medium sized machines are being installed in large numbers for feeding the local grid notably in the state of California (where over 10,000 medium sized wind generators have been installed in little more than 3 years for feeding the grid) and in Denmark. Fig. 114 shows a typical modern 55kW, 15m diameter Windmatic wind turbine, from Denmark. Machines of this size may in future be of considerable relevance for larger scale irrigation pumping than is feasible with more traditional mechanical windpumps, (see Gilmore et al [43], and Nelson et al [44]).

ii.   State-of-the-Art

There are two distinct end-uses for windpumps, namely either irrigation or water supply, and these give rise to two distinct categories of windpump because the technical, operational and economic requirements are generally different for these end uses. That is not to say that a water

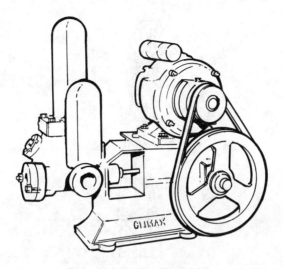

Fig. 106   Electric motor powered, belt driven piston pump (Climax) (note air chambers provided to prevent water hammer)

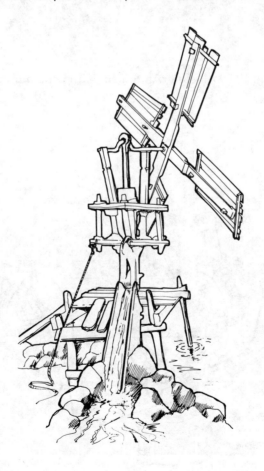

Fig. 107   Wooden indigenous windmill pump for pumping sea water into salt pans on the Island of Sal, Cape Verde

Fig. 108 All-steel 'American' farm wind pump

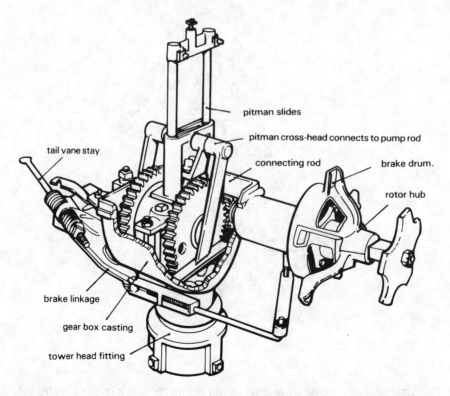

Fig. 109 Gearbox from a typical back-geared 'American' farm windmill

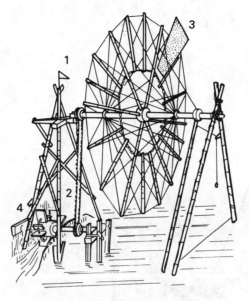

Fig. 110 Chinese chain windmill
   1 sail
   2 chain drive
   3 flag indicating wind direction
   4 paddle pump
(note only one sail shown fitted for clarity (after [51])

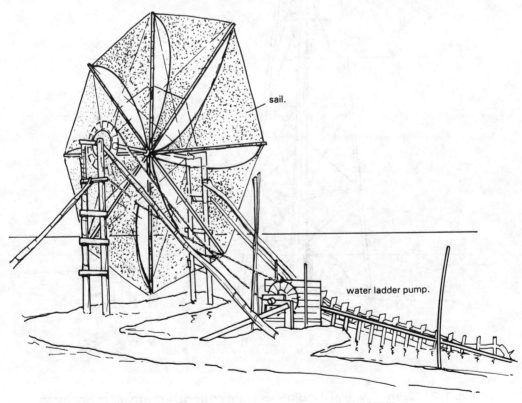

Fig. 111 Thai windpump (after Schioler [24])

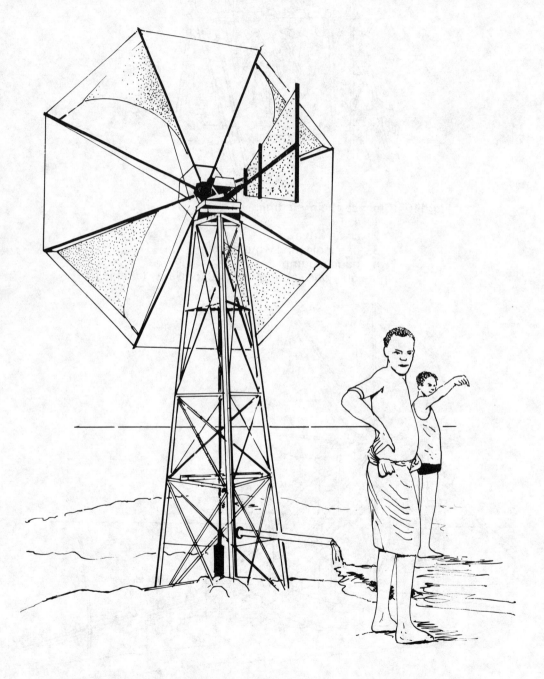

Fig. 112 'Cretan' type of windmill used on an irrigation project in Southern Ethiopia (after Fraenkel [15])

supply windpump cannot be used for irrigation (they quite often are) but irrigation designs are generally unsuitable for water supply duties.

Most water supply windpumps must be ultra-reliable, to run unattended for most of the time (so they need automatic devices to prevent overspeeding in storms), and they also need the minimum of maintenance and attention and to be capable of pumping water generally from depths of 10m or more. A typical farm windpump should run for over 20 years with maintenance only once every year, and without any major replacements; this is a very demanding technical requirement since typically such a wind pump must average over 80,000 operating hours before anything significant wears out; this is four to ten times the operating life of most small diesel engines or about 20 times the life of a small engine pump. Windpumps to this standard therefore are usually industrially manufactured from steel components and drive piston pumps via reciprocating pump rods. Inevitably they are quite expensive in relation to their power output, because of the robust nature of their construction. But American, Australian and Argentinian ranchers have found the price worth paying for windpumps that achieve high reliability and minimum need for human intervention, as this is their main advantage over practically any other form of pumping systems.

Irrigation duties on the other hand are seasonal (so the windmill may only be useful for a limited fraction of the year), they involve pumping much larger volumes of water through a low head, and the intrinsic value of the water is low. Therefore any windpump developed for irrigation has to be low in cost and this requirement tends to overide most other considerations. Since irrigation generally involves the farmer and/or other workers being present, it is not so critical to have a machine capable of running unattended. Therefore windmills used for irrigation in the past tend to be indigenous designs that are often improvized or built by the farmer as a method of low-cost mechanization; (eg. Figs. 110, 111 and 112). If standard farm windpumps (Fig. 108) are used for irrigation, usually at much lower heads than are normal for water supply duties, there are quite often difficulties in providing a piston pump of sufficient diameter to give an adequate swept volume to absorb the power from the windmill. Also most farm windpumps have to be located directly over the pump, on reinforced concrete foundations, which usually limits these machines to pumping from wells or boreholes rather than from open water. A suction pump can be used on farm windmills with suction heads of up to about 5-6m from surface water; (see Fig. 115 for typical farm windpump installation configurations). Most indigenous irrigation windpumps, on the other hand, such as those in China, use rotary pumps of one kind or another which are more suitable for low heads; they also do not experience such high mechanical forces as an industrial windpump, (many of which lift their pump rods with a pull of over 1tonne, quite enough to "uproot" any carelessly installed pump).

Attempts have been made recently to develop lower cost steel windpumps that incorporate the virtues of the heavier older designs. Most farm-windpumps, even though still in commercial production, date back to the 1920s or earlier and are therefore unecessarily heavy and expensive to manufacture, and difficult to install properly in remote areas. Recently various efforts have been made to revise the traditional farm windpump concept into a lighter and simpler modern form. Fig. 116 shows the "IT Windpump", which is half the weight of most traditional farm windpump designs of a similar size, and is manufactured in Kenya as the "Kijito" and in Pakistan as the "Tawana". The latter costs only about half as much as American or

Australian machines of similar capability. It is possible therefore that through developments of this kind, costs might be kept low enough to allow the marketing of all steel windpumps that are both durable like the traditional designs, yet cheap enough to be economic for irrigation.

4.7.2   Principles of Wind Energy Conversion

i.   Power available in the wind

The power in the wind is proportional to the wind speed cubed; the general formula for power in the wind is:

$$P = \tfrac{1}{2}\rho A V^3$$

where P is the power available in watts, ρ is the density of air (which is approximately $1.2 kg/m^3$ at sea level), A is the cross-section (or swept area of a windmill rotor) of air flow of interest and V is the instantaneous free-stream wind velocity. If the velocity, V, is in m/s (note that 1m/s is almost exactly 2 knots or nautical miles per hour), the power in the wind at sea level is:

$$P = 0.6V^3 \text{ watts/m}^2 \text{ of rotor area}$$

Because of this cubic relationship, the power availability is extremely sensitive to wind speed; doubling the wind speed increases the power availability by a factor of eight; Table 16 indicates this variability.

Table 16   POWER IN THE WIND AS A FUNCTION OF WIND SPEED
IN UNITS OF POWER PER UNIT AREA OF WIND STREAM

wind speed	m/s	2.5	5	7.5	10	15	20	30	40
	km/h	9	18	27	36	54	72	108	144
	mph	6	11	17	22	34	45	67	90
power density	kW/m^2	.01	.08	.27	.64	2.2	5.1	17	41
	hp/ft^2	.001	.009	.035	.076	.23	.65	2.1	5.2

This indicates the very high variability of wind power, from around $10W/m^2$ in a light breeze up to $41,000 Wm^2$ in a hurricane blowing at 144km/h. This extreme variability greatly influences virtually all aspects of system design. It makes it impossible to consider trying to use winds of less than about 2.5m/s since the power available is too diffuse, while it becomes essential to shed power and even shut a windmill down if the wind speed exceeds about 10-15m/s (25-30mph) as excessive power then becomes available which would damage the average windmill if it operated under such conditions.

The power in the wind is a function of the air-density, so it declines with altitude as the air thins, as indicated in Table 17.

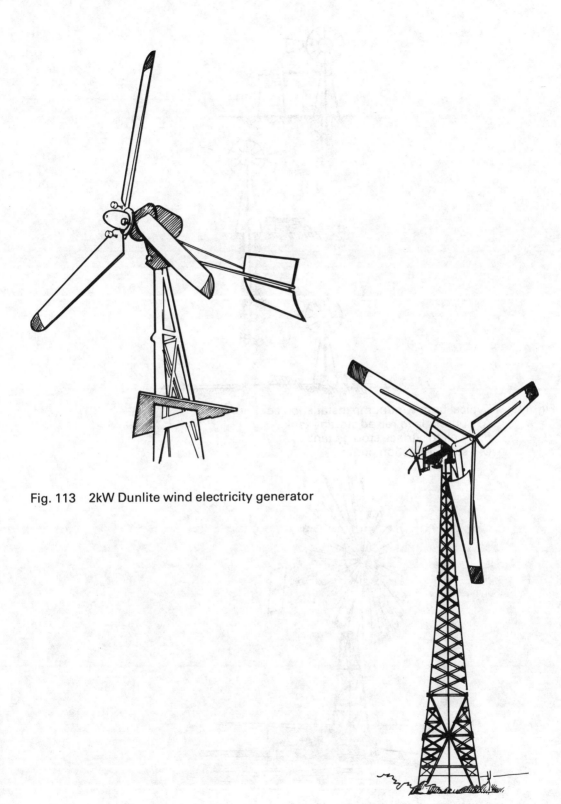

Fig. 113 2kW Dunlite wind electricity generator

Fig. 114 55kW Windamatic wind electricity generator

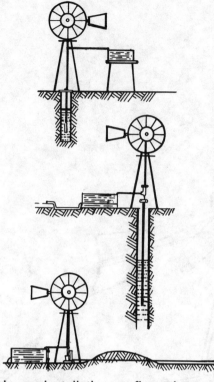

Fig. 115 Typical farm windpump installation configurations
  A. borehole to raised storage tank
  B. well to surface storage tank
  C. surface suction pump

Fig. 116 IT windpump, made in Kenya as the 'Kijito' and in Pakistan as the 'Tawana'

Table 17   VARIATION OF AIR DENSITY WITH ALTITUDE

altitude a.s.l.	(ft)	0	2,500	5,000	7,500	10,000
	(m)	0	760	1,520	2,290	3,050
density correction factor		1.00	0.91	0.83	0.76	0.69

Because the power in the wind is so much more sensitive to velocity rather than to air density, the effect of altitude is relatively small. For example the power density of a 5m/s wind at sea level is about 75 watts/m²; however, due to the cube law, it only needs a wind speed of 5.64m/s at 3,000m a.s.l. to obtain exactly the same power of 75 watts/m². Therefore the drop in density can be compensated for by quite a marginal increase in wind velocity at high altitudes.

ii.   Energy available in the wind

Because the speed of the wind constantly fluctuates, its power also varies to a proportionately greater extent because of the cube law. The energy available is the summed total of the power over a given time period. This is a complex subject (Lysen [45] gives a good introduction to it). The usual starting point to estimate the energy available in the wind at a specific location is some knowledge of the mean or average wind speed over some predefined time period; typically monthly means may be used. The most important point of general interest is that the actual energy available from the wind during a certain period is considerably more than if you take the energy that would be produced if the wind blew at its mean speed without variation for the same period. Typically the energy available will be about double the value obtained simply by multiplying the instantaneous power in the wind that would correspond to the mean wind speed blowing continuously, by the time interval. This is because the fluctuations in wind speed result in the average power being about double that which occurs instantaneously at the mean wind speed. The actual factor by which the average power exceeds the instantaneous power corresponding to the mean windspeed can vary from around 1.5 to 3 and depends on the local wind regime's actual variability. The greater the variability the greater this factor.

However, for any specific wind regime, the energy available will still generally be proportional to the mean wind speed cubed. We shall discuss later in this section how to determine the useful energy that can be obtained from a wind regime with respect to a particular windmill.

iii.   Converting wind power to shaft power

There are two main mechanisms for converting the kinetic energy of the wind into mechanical work; both depend on slowing the wind and thereby extracting kinetic energy. The crudest, and least efficient technique is to use drag; drag is developed simply by obstructing the wind and creating turbulence and the drag force acts in the same direction as the wind. Some of the earliest and crudest types of wind machine, known generically as "panamones", depend on exposing a flat area on one side of a rotor to the wind while shielding (or reefing the sails) on the other side; the resulting differential drag force turns the rotor.

The other method, used for all the more efficient types of windmill, is to produce lift. Lift is produced when a sail or a flat surface is mounted at a small angle to the wind; this slightly deflects the wind and produces a large force perpendicular to the direction of the wind with a much smaller drag force. It is this principle by which a sailing ship can tack at speeds greater than the wind. Lift mainly deflects the wind and extracts kinetic energy with little turbulence, so it is therefore a more efficient method of extracting energy from the wind than drag.

It should be noted that the theoretical maximum fraction of the kinetic energy in the wind that could be utilized by a "perfect" wind turbine is approximately 60%. This is because it is impossible to stop the wind completely, which limits the percentage of kinetic energy that can be extracted.

iv. Horizontal and vertical axis rotors

Windmills rotate about either a vertical or a horizontal axis. All the windmills illustrated so far, and most in practical use today, are horizontal axis, but research is in progress to develop vertical axis machines. These have the advantage that they do not need to be orientated to face the wind, since they present the same cross section to the wind from any direction; however this is also a disadvantage as under storm conditions you cannot turn a vertical axis rotor away from the wind to reduce the wind loadings on it.

There are three main types of vertical axis windmill. Panamone differential drag devices (mentioned earlier), the Savonius rotor or "S" rotor (Fig. 117) and the Darrieus wind turbine (Fig. 118). The Savonius rotor consists of two or sometimes three curved interlocking plates grouped around a central shaft between two end caps; it works by a mixture of differential drag and lift. The Savonius rotor has been promoted as a device that can be readily improvised on a self-build basis, but its apparent simplicity is more perceived than real as there are serious problems in mounting the inevitably heavy rotor securely in bearings and in coupling its vertical drive shaft to a positive displacement pump (it turns too slowly to be useful for a centrifugal pump). However the main disadvantages of the Savonius rotor are two-fold:

a. it is inefficient, and involves a lot of construction material relative to its size, so it is less cost-effective as a rotor than most other types;

b. it is difficult to protect it from over-speeding in a storm and flying to pieces.

The Darrieus wind turbine has airfoil cross-section blades (streamlined lifting surfaces like the wings of an aircraft). These could be straight, giving the machine an "H"-shaped profile, but in practice most machines have the curved "egg-beater" or troposkien profile as illustrated. The main reason for this shape is because the centrifugal force caused by rotation would tend to bend straight blades, but the skipping rope or troposkien shape taken up by the curved blades can resist the bending forces

Fig. 117 Savonius Rotor vertical-axis windpump in Ethiopia. It was found to be less cost-effective than the 'Cretan' windmill of Fig. 112 (See ref. [15])

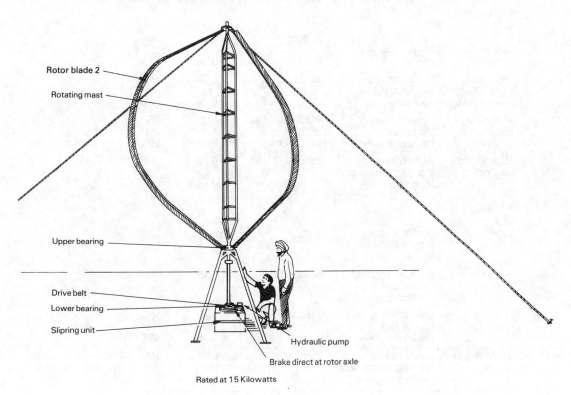

Fig. 118 Typical Troposkien shaped Darrieus vertical axis wind turbine

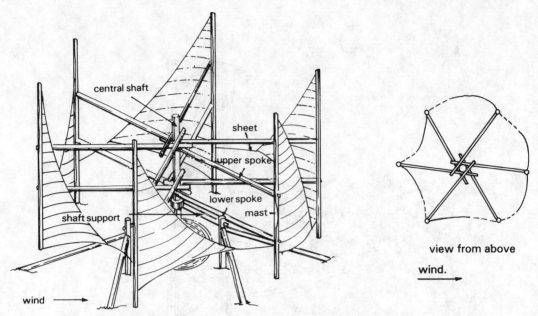

Fig. 119 Turks and Caicos islands vertical-axis sail rotor (after UNESCAP [51])

Table 18 COMPARISON BETWEEN DIFFERENT ROTOR TYPES

Type	Performance characteristic	Manufacturing requirements	Cp	Solidarity σ	t.s.r.* (Optimum)
Horizontal axis —					
Cretan sail or flat paddles	Medium starting torque and low speed	Simple	0.05 to 0.15	50% 1.5-2.0	1.5-2.0
Cambered plate fan (American)	High starting torque and low speed	Moderate	0.15 to 0.30	50 to 80%	1-1.5
Moderate speed aerogenerator	Low starting torque and moderate speed	Moderate, but with some precision	0.20 to 0.35	5 to 10%	3-5
High speed aero-gen.	Almost zero starting torque and high speeds	Precise	0.30 to 0.45	under 5%	5-10
Vertical axis —					
Panemone	Medium starting torque and low speed	Simple	under 0.10	50%	.4-.8
Savonius rotor	Medium starting torque and moderate speed	Moderate	0.15	100%	.8-1
Darrieus rotor	Zero starting torque and moderate speed	Precise	0.25 to 0.35	10% to 20%	3-5
VGVAWT or Gyromill	Zero or small starting torque and moderate speed	Precise	0.20 to 0.35	15% to 40%	2-3

*t.s.r. = tip-speed ratio (λ)

effectively. Darrieus-type vertical axis turbines are quite efficient, since they depend purely on lift forces produced as the blades cross the wind (they travel at 3 to 5 times the speed of the wind, so that the wind meets the blade at a shallow enough angle to produce lift rather than drag). The Darrieus was predated by a much cruder vertical axis windmill with Bermuda (triangular) rig sails from the Turks and Caicos Islands of the West Indies (Fig. 119). This helps to show the principle by which the Darrieus works, because it is easy to imagine the sails of a Bermuda rig producing a propelling force as they cut across the wind in the same way as a sailing yacht; the Darrieus works on exactly the same principle.

There are also two main types of Darrieus wind turbine which have straight blades; both control overspeed and consequent damage to the blades by incorporating a mechanism which reefs the blades at high speeds. These are the Variable Geometry Vertical Axis Wind Turbine (VGVAWT) developed by Musgrove in the UK and the Gyromill Variable Pitch Vertical Axis Wind Turbine (VPVAWT), developed by Pinson in the USA. Although the Musgrove VGVAWT has been tried as a windpump by P I Engineering, all the current development effort is being channelled into developing medium to large electricity grid-feeding, vertical-axis wind-generators, of little relevance to irrigation pumping.

Vertical axis windmills are rarely applied for practical purposes, although they are a popular subject for research. The main justification given for developing them is that they have some prospect of being simpler than horizontal axis windmills and therefore they may become more cost-effective. This still remains to be proved.

Most horizontal axis rotors work by lift forces generated when "propeller" or airscrew like blades are set at such an angle that at their optimum speed of rotation they make a small angle with the wind and generate lift forces in a tangential direction. Because the rotor tips travel faster than the roots, they "feel" the wind at a shallower angle and therefore an efficient horizontal axis rotor requires the blades to be twisted so that the angle with which they meet the wind is constant from root to tip. The blades or sails of slow speed machines can be quite crude (as in Fig. 107) but for higher speed machines they must be accurately shaped airfoils (Figs. 113 & 114); but in all three examples illustrated, the principle of operation is identical.

v. <u>Efficiency, power and torque characteristics</u>

Any wind turbine or windmill rotor can be characterized by plotting experimentally derived curves of power against rotational speed at various windspeeds; Fig. 120 A. Similarly the torque produced by a wind rotor produces a set of curves such as in Fig. 120 B.

The maximum efficiency coincides with the maximum power output in a given windspeed. Efficiency is usually presented as a non-dimensional ratio of shaft-power divided by wind-power passing through a disc or shape having the same area as the vertical profile of the windmill rotor; this ratio is known as the "Power Coefficient" or C_p and is numerically expressed as:

$$C_p = \frac{P}{\frac{1}{2}\rho AV^3}$$

the speed is also conventionally expressed non-dimensionally as the "tip-speed ratio" (λ). This is the ratio of the speed of the windmill rotor tip, at radius R when rotating at ω radians/second, to the speed of the wind, V, and is numerically:

$$\lambda = \frac{\omega R}{V}$$

When the windmill rotor is stationery, its tip-speed ratio is also zero, and the rotor is stalled. This occurs when the torque produced by the wind is below the level needed to overcome the resistance of the load. A tip-speed ratio of 1 means the blade tips are moving at the same speed as the wind (so the wind angle "seen" by the blades will be 45°) and when it is 2, the tips are moving at twice the speed of the wind, and so on.

The C_p versus curves for three different types of rotor, with configurations A, B, C, D, E1, E2 and F as indicated, are shown in Fig. 121. The second set of curves show the torque coefficients, which are a non-dimensional measure of the torque produced by a given size of rotor in a given wind speed (torque is the twisting force on the drive shaft). The torque coefficient, C_t, is defined as:

$$C_t = \frac{T}{\frac{1}{2} \rho A V^2 R}$$

where T is the actual torque at windspeed V for a rotor of that configuration and radius R.

vi. **Rotor solidity**

"Solidity" (σ) is a fairly graphic term for the proportion of a windmill rotor's swept area that is filled with solid blades. It is generally defined as the ratio of the sum of the width, or "chords" of all the blades to the circumference of the rotor; i.e. 24 blades with a chord length (leading edge to trailing edge) of 0.3m on a 6m diameter rotor would have a tip solidity of:

$$\sigma = \frac{(24 \times 0.3)}{6\pi}$$

Multi-bladed rotors, as used on windpumps, (eg. rotor "B" in Fig. 121) are said to have high "solidity", because a large proportion of the rotor swept area is "solid" with blades. Such machines have to run at relatively low speeds and will therefore have their blades set at quite a coarse angle to the plain of rotation, like a screw with a coarse thread. This gives it a low tip-speed ratio at its maximum efficiency, of around 1.25, and a slightly lower maximum coefficient of performance than the faster types of rotor such as "D", "E" and "F" in the figure. However, the multi-bladed rotor has a very much higher torque coefficient at zero tip-speed ratio (between 0.5 and 0.6) than any of the other types. Its high starting torque (which is higher than its running torque) combined with its slow speed of rotation in a given wind make it well-suited to driving reciprocating borehole pumps.

In contrast, the two or three-bladed, low-solidity, rotors "E1" and

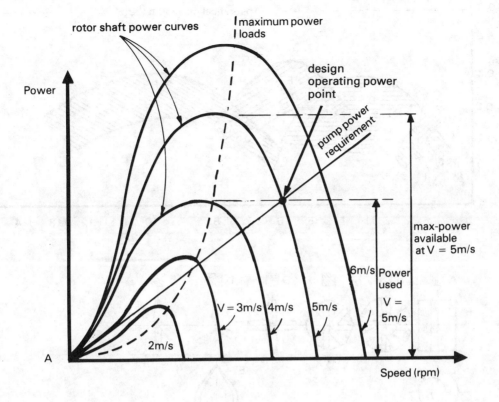
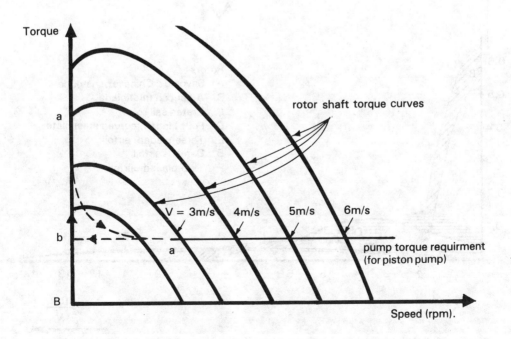

Fig. 120 The power (A) and torque (B) of a wind rotor as a function of rotational speed for difference wind speeds

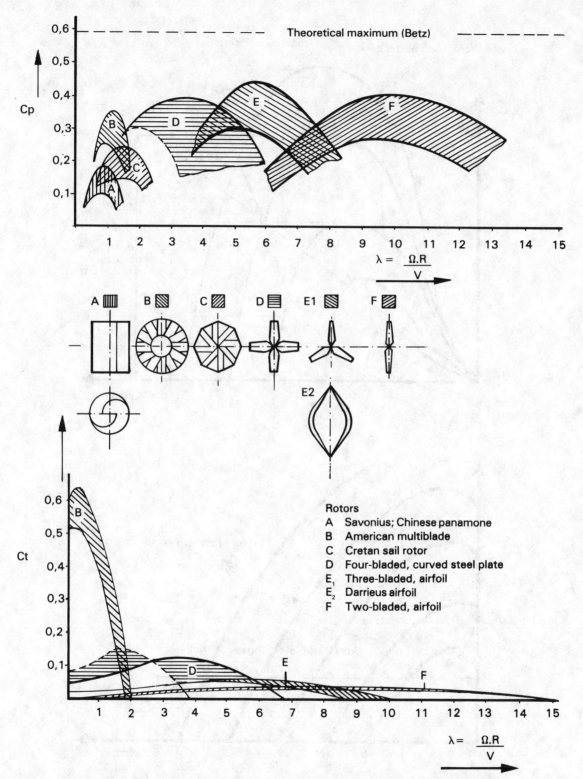

Fig. 121 The power coefficients (Cp) (above) and the torque coefficients (Ct) of various types of wind turbine rotor plotted against tip-speed ratio (λ) (after Lysen/CWD [45])

"F" in Fig. 121, are the most efficient, (with the highest values for C_p), but their tips must travel at six to ten times the speed of the wind to achieve their best efficiency. To do so they will be set at a slight angle to the plain of rotation, like a screw with a fine thread and will therefore spin much faster for a given windspeed and rotor diameter than a high solidity rotor. They also have very little starting torque, almost none at all, which means they can only start against loads which require little torque to start them, like electricity generators (or centrifugal pumps) rather than positive displacement pumps.

All this may sound academic, but it is fundamental to the design of wind rotors; it means that multi-bladed "high-solidity" rotors run at slow speeds and are somewhat less efficient than few-bladed "low solidity" rotors, but they have typically five to twenty times the starting torque.

vii. <u>Matching rotors to pumps</u>

High solidity rotors are typically used in conjunction with positive displacement (piston) pumps, because, as explained in Section 3.5, single-acting piston pumps need about three times as much torque to start them as to keep them going. Low solidity rotors, on the other hand, are best for use with electricity generators or centrifugal pumps, or even ladder pumps and chain and washer pumps, where the torque needed by the pump for starting is less than that needed for running at design speed. Table 18 indicates the relative characteristics and C_p values for various typical wind rotor types so far described.

Fig. 120 A and B shows the load lines for a positive displacement direct-driven pump superimposed on the wind rotor output curves. The dotted line on Fig. 120 A indicates the locus of the points of maximum power; the system will only function continuously when the operating point is to the right of the line of maximum power, as under that condition any slight drop in wind speed causes the machine to slow and the power absorbed by the shaft to increase, which results in stable operation. The operating point can only remain to the left of the maximum power locus under conditions of increasing windspeed. It can be seen that the positive displacement pump requires more or less constant torque of 10Nm in the example, once rotation has been established, but it needs over three times as much torque to start it for reasons explained in Section 3.5. The torque curves in Fig. 120 B indicate that 5m/s windspeed is needed to produce the torque required to start the windpump rotating, but once rotation has commenced, the windspeed can fall to 3m/s before the operating point moves to the left of th maximum powere locus and the windpump will stop. Note that the broken line a'-a represents a transient condition that only occurs momentarily when the windpump starts to rotate.

To extract the maximum power from a windpump at all times would require a load which causes the operating point to follow close to the locus of maximum power; (Fig.120). The figure also indicates that that the operating point will always be where the windpump rotor curve for the windspeed prevailing at a given moment coincides with the pump load line. In the example, the operating point is shown for a windspeed of 5m/s; in this example, it can be seen that only about two-thirds of the maximum power that could be produced in this wind speed is used by the pump, because its load line diverges from the cubic maximum power curve. This discrepancy is a mis-match between the prime-mover (the windmill rotor) and the load (the pump). The proportion of the power available from the rotor in a given windspeed which is usefully applied is known as the "matching efficiency", and is analysed in detail in Pinilla, et al [46]. The figure illustrates how this mis-match becomes progressively worse as the wind speed increases. This mis-match is actually less serious than it may seem, since the time when the best efficiency is needed is at low windspeeds when, fortunately, the best efficiency is achieved. When a windmill is running fast enough to be badly matched with its pump, it means that the wind is blowing more strongly than usual and the chances are that the output, although theoretically reduced by bad matching, will be more than adequate, as the extra speed will compensate for the reduction in efficiency.

It may be thought that centrifugal pumps would match better with a windmill than positive displacement pumps, but in practice their efficiency falls rapidly to zero below a certain threshold running speed at a fixed static head. In otherwords, centrifugal pumps do not readily run with adequate efficiency over as wide a speed range as is necessary to match most windmills rotors and they are therefore not generally used with windmills (except with intermediate electrical transmission which can modify the relationship between the pump and windmill speeds).

When generators are used as a load, instead of pumps, a much better match can be obtained. Wind generators therefore tend to have a better matching efficiency over their whole range of operating speeds than windpumps; the interested reader is referred to a text on this subject, such as Lysen [45].

There is considerable scope for improving the overall performance of wind pumps by developing methods of improving the rotor-to-pump match over a wider range of windspeeds; a certain amount of work is being carried out in this field and if successful could result in considerably more effective windpumps in the future. But in the meantime the main problem is to choose the most appropriate pump size for a given windmill in a given wind regime and location. Fig. 122 shows how the pump load line can be altered simply by changing the mean pump rod pull, either by changing the stroke (by lengthening or shortening the crank) or by changing the diameter of the pump being used. A longer stroke and/or a larger pump will increase the pump rod force, and increase the mean torque requirement and hence the slope of the load line, and vice-versa. In Fig. 122 it is clear that increasing the load increases the hydraulic output at higher speeds, but it also increases the value of V_s, the starting windspeed. Therefore, pump "C" in the diagram will start in a much lighter wind than the other pumps, but because of the shallower load line the output will be much smaller in high winds. There is therefore an important trade-off between achieving starting in adequately light winds and achieving a good output.

The operating characteristic of a typical windpump, given in Fig. 123, shows how if the start-up windspeed is V_s, a windpump can run down to a slightly lower windspeed V_{min} (as explained earlier, assuming the use of a piston pump). It reaches its best match with the rotor at windspeeds close to V_{min} (in theory at $0.8V_s$) [46] which is the "Design Windspeed", and then increases its output almost linearly with windspeed to V_r (its rated windspeed). At still higher windspeeds means must be introduced to prevent it speeding up further, or the machine may be over-loaded and damaged or destroyed; various methods for doing this are discussed in the next section

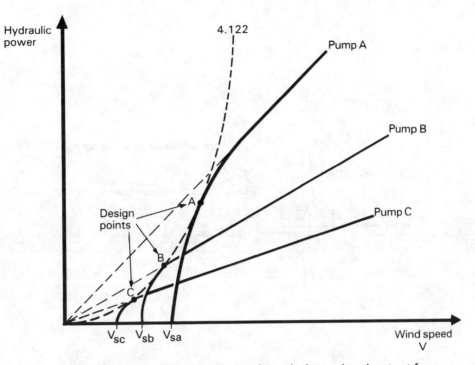

Fig. 122 The trade-off between starting windspeed and output for differently loaded windpumps

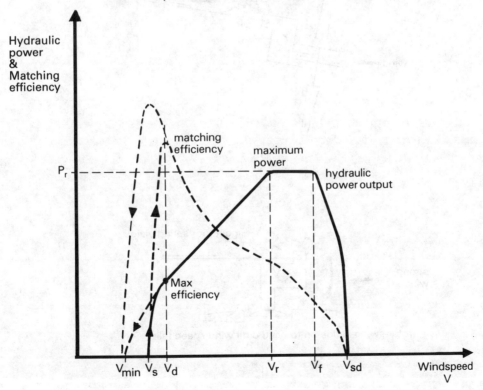

Fig. 123 The operating characteristic of a windpump showing how the power output and matching efficiency vary with windspeed

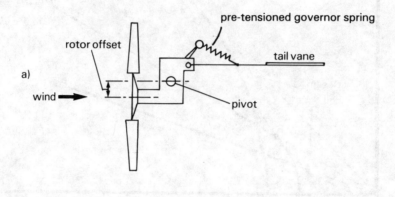

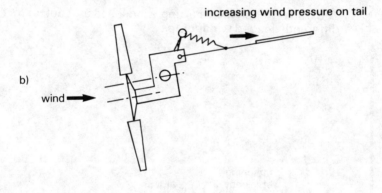

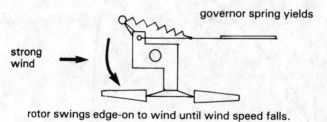

Fig. 124 Typical windpump storm protection method in which rotor is yawed edge-on to the wind (plan view)

below. At very high windspeeds, the only safe course of action is to make the windmill "reef", "furl" or "shut-down"; the figure shows how this process commences at a windspeed V_f (furling speed) and is completed at windspeed V_{sd} (shut-down).

viii.   Methods of storm protection and furling

Windmills must have a means to limit the power they can deliver, or else they would have to be built excessively strongly (and expensively) merely to withstand only occasional high power outputs in storms. Sailing ships "take in canvas" by wholly or partially furling the sails (manually) when the wind is too strong, and Cretan sail windmills and other such simple traditional designs generally use exactly the same technique; fewer sails are used in high winds or else the sails are partially rolled around their spars. Metal farm windmills, however, have fixed steel blades, so the solution most generally adopted is to mount the rotor offset from the tower centre (Fig. 124) so that the wind constantly seeks to turn the rotor behind the tower. Under normal conditions the rotor is held into the wind by a long tail with a vane on it. This vane is hinged, and fixed in place with a pre-loaded spring (as illustrated), then when the wind load on the rotor reaches a level where the force is sufficient to overcome the pre-tension in the spring, the tail will start to fold until the wind pushes the rotor around so that it presents its edge to the wind, as in Fig. 124. This furling process starts when the rated output is reached and if the windspeed continues to rise, it increases progressively until the machine is fully furled. Then when the wind drops, the spring causes the tail vane to unfold again and turn the rotor once again to face the wind. On commercial farm windmills, this action is normally completely automatic.

Wind-generators and other windturbines with high speed, low-solidity rotors often use a mechanism which changes the blade pitch; e.g. the Dunlite machine of Fig. 113 which has small counter-weights, visible near the rotor hub, which force the blades into a coarser pitch under the influence of centrifugal force when the rotor reaches its furling speed, against the force of a spring enclosed in the hub. Alternatively air-brake flaps are deployed to prevent overspeed. Larger windturbines do not use tail vanes to keep them facing the wind, as they cannot stand being yawed as fast as might occur if there is a sudden change in wind direction. Instead they usually have a worm-reduction gear mechanism similar to that in a crane, which inches them round to face the wind; this can be electrically powered on signals from a small wind direction vane, or it can use the mechanism visible on the Windmatic in Fig. 114, used on large windmills for several centuries, where a sideways mounted windrotor drives the orientating mechanism every time the main rotor is at an angle other than at right angles to the wind direction.

4.7.3   The Wind Resource

It is not intended to attempt a detailed discussion of the complex subject of the causes and behaviour of the wind or how it is measured and analysed; useful references are Lysen [45], Park [46], Golding [47] and the W.M.O. [48].

The main fact to be aware of is that although the wind is extremely variable and unpredictable on a minute-by-minute or an hour-by-hour basis, the actual average windspeed at a given location over any given month of the year will not differ much from year to year. So if mean monthly windspeed figures are available, taken over a number of years, a reasonable prediction of the performance of a windpump should be possible. A variety of methods may be used to do this, some of which are described in the next section.

i.   The minimum wind requirements for windpumps

It has shown that, typically, windpumps require an average "least windy month" windspeed of about 2.5m/s to begin to be economically competitive, (eg. Fraenkel, [40]). Because of the cube relationship between windspeed and energy availability, which is true for any optimally matched wind pump, and wind regime, the economics of windpumps are very sensitive to windspeed. Therefore, windpumps are one of the most cost-effective options (compared with engines or any other prime-movers) for pumping in locations with mean wind speeds exceeding about 4m/s, but, conversely, they are not at all cost-competitive where mean windspeeds are significantly below 2.5m/s. Fig 125 rather crudely indicates the world's windspeed distribution pattern. This is considerably simplified and a much more detailed treatment of this topic is available from the World Meteorological Organization; [48]. It can be seen that much of the world, with the exception of the centres of the major land-masses, and the equatorial forested regions, is suitable for deploying windpumps. As a rule of thumb, areas that are free of trees (i.e. savannah grasslands, semi-deserts and deserts) tend to be windy and suitable for windpumps, while forested and wooded areas not only have less wind but trees make siting of windmills difficult unless very high towers can be used.

Various studies on the potential market for windmills in different parts of the world, eg. [49], plus numerous country-specific studies of meteorological data, suggest that quite a number of developing countries probably have areas with adequate wind speeds for the use of windpumps. Some of these are listed below; those where windpumps are already known to be in at least moderately widespread use are marked (+):

Algeria, Argentina(+), Cape Verde Islands(+), Chile, China (+), Cyprus (+), Ecuador, Egypt, Ethiopia, India, Jordan, Kenya(+), Libya, Madagascar, Malta, Mauritania, Mauritius, Morocco (+), Mozambique, Namibia(+), Oman, Pakistan, Peru(+), Senegal, Somalia, Sudan, Syria, Tanzania, Thailand(+), Tunisia(+), Uruguay, Zambia, Zimbabwe(+).

There are also many small islands which are not listed but which invariably have adequate wind regimes due to the proximity of the open ocean.

ii.   Variation of wind speed with height

The speed of the wind increases with height. The rate of increase is dependent partly on the height and partly on the nature of the ground surface. This is because rough ground, with many uneven trees, bushes or buildings, causes turbulence, while a flat and unobstructed surface like the sea or a flat grassy plain allows the air to flow smoothly which results in higher windspeeds nearer to the surface. The relationship between windspeed and height can be estimated as follows:

$$\frac{V}{V_r} = \left(\frac{H}{H_r}\right)^a$$

where V is the wind velocity at height H and V_r is a reference wind

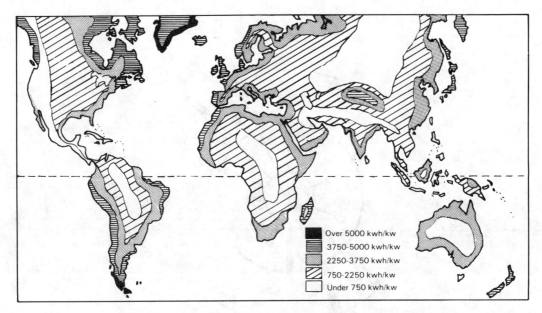

Fig. 125 Annual mean wind speeds (approximate indication)

Location		Average wind speed in m/s													Years of observation
		Jan	Feb	Mar	Apr	May	Jun	Jul	Aug	Sep	Oct	Nov	Dec	Year	
28° 37'N, 77° 13'E	New Delhi	2.2	2.8	3.1	3.1	3.6	4.2	2.8	2.5	2.8	1.7	1.9	2.2	2.7	
29° 27'N, 79° 41'E	Mukeshwar	3.1	3.4	3.6	4.1	4.6	4.3	3.4	2.8	2.8	2.8	2.9	2.9	3.4	
22° 30'N, 88° 20'E	Calcutta	1.1	1.4	2.3	3.6	4.5	3.4	3.2	2.8	2.3	1.4	1.1	1.0	2.4	
24° 48'N, 85° 00'E	Gaya	1.7	1.9	2.5	2.8	3.3	3.3	3.1	3.1	2.5	1.7	1.4	1.4	2.4	
19° 49'N, 85° 54'E	Puri	3.3	4.4	5.7	6.7	7.3	6.5	6.5	5.5	4.4	3.4	2.8	2.9	4.9	
19° 20'N, 85° 00'E	Gupalpur	2.8	3.6	5.0	6.4	6.9	4.7	4.7	4.2	3.3	3.1	3.1	2.8	4.2	
27° 06'N, 72° 22'E	Phaloch	2.8	2.4	3.6	3.9	5.7	7.1	6.6	5.4	4.6	3.2	3.3	2.3	4.2	1931
22° 42'N, 75° 54'E	Indore	2.8	3.0	3.6	4.3	6.8	7.5	7.3	6.0	5.1	2.7	2.1	2.0	4.4	to
17° 45'N, 73° 07'E	Harnai	3.9	3.9	4.2	4.7	4.7	4.4	6.1	5.8	3.6	2.8	3.1	3.1	4.2	1960
16° 24'N, 73° 25'E	Devgarh	3.6	3.9	4.2	4.7	5.0	5.8	8.1	7.2	4.2	3.1	3.1	3.1	4.7	
22° 18'N, 70° 53'E	Rajkot	3.6	3.9	5.3	5.6	7.2	7.8	7.8	6.4	4.7	3.3	3.1	3.1	5.1	
22° 48'N, 74° 18'E	Dohad	2.5	2.8	3.6	4.7	7.2	8.3	8.1	6.1	3.1	2.5	1.9	1.9	4.4	
17° 22'N, 78° 26'E	Hyderabad	2.2	2.5	2.8	3.1	3.3	6.7	6.1	5.0	3.6	2.5	2.2	1.9	3.5	
15° 54'N, 74° 36'E	Belgaum	3.1	3.3	3.3	3.9	5.3	6.7	7.2	6.4	4.7	2.8	3.1	3.3	4.4	
11° 00'N, 76° 54'E	Coimbatore	2.9	3.0	3.3	4.1	6.4	9.1	8.6	8.6	5.6	4.5	2.6	2.8	5.1	
10° 50'N, 78° 43'E	Tiruchirapalli	2.8	2.2	2.5	2.8	4.7	8.1	9.7	7.2	5.3	3.1	2.5	3.1	4.5	

Source: "Climatological tables of observations in India" given in a programme for the Development and Utilization of windmills in India (National Aeronautical Laboratory Bangalore September 1978)

Fig. 126 Typical presentation of long term wind data as monthly averages

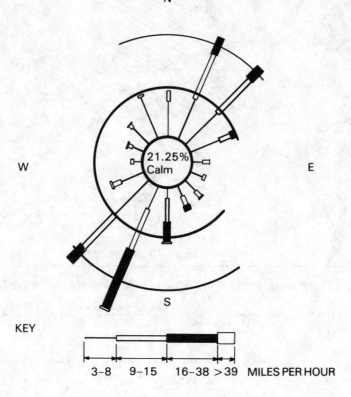

Fig. 127 Wind rose

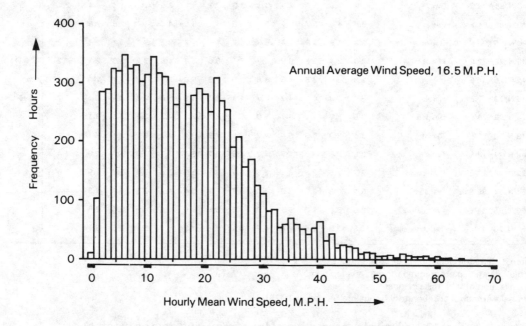

Fig. 128 Wind velocity-frequency histogram

velocity measured at height H_r. The exponent "a" is a function of the surface roughness, as follows, [50]:

type of terrain	a
smooth sea or sand	0.10
low grass steppe	0.13
high grass and small bushes	0.19
woodlands and urban areas	0.32

For example, if there is high grass and small bushes, and a mean reference windspeed, for example, of 5m/s recorded at the standard meteorological office recommended height of 10m, this can be adjusted to obtain the mean windspeed at 20m windmill hub height as follows:

$$\left(\frac{V}{5}\right) = \left(\frac{H}{10}\right)^{0.19}$$

from which V= 5.7m/s   A gain of 0.7m/s from mounting a windmill at 20m rather than at 10m may sound small in relation to the cost of the extra high tower required, but the energy available at those two heights will be related to the cube of the velocities (assuming optimally matched pumps in each case) and will therefore be:

$$\left(\frac{5.7}{5.0}\right)^3 = 1.48$$

this shows that a 48% increase in energy availability can be gained in terrain of that type from using a 20m tower instead of a 10m tower (or a windmill with a smaller rotor could be used to gain the same energy - in this case the rotor area could be reduced so that a windmill with 20% smaller rotor diameter on a 20m tower would be used compared with one on a 10m tower).

iii. Effects of obstructions

Any obstruction to the wind has a wake extending up to 20 or 30 diameters (of the obstruction) downwind. The wake is depleted of wind energy compared with the surrounding wind, and is turbulent. For example a large mango tree (or similar rounded, well-leafed tall tree) can have a wake which even 200-300m downwind has 10% less wind energy than either side of it. Sharp edged and irregular obstructions such as rock outcrops, cliffs and escarpments, or large buildings can cause violent turbulence which, apart from depleting the energy available, can cause damage to a windmill located nearby.

Therefore it is normal to recommend that windmills are mounted so that the rotor is at least 200-300m from any significant obstruction to the wind. Ideally if obstructions like trees or buildings are nearby, the rotor should be mounted on a high enough tower so its bottom edge is a clear 5m or more above the highest point of the obstruction. In reality it is often impossible to avoid obstructions, so the least that can be done is to try and locate the windmill so that it is unobstructed from the direction of the prevailing wind. Because the wind lifts to go over an obstruction, siting windmills with large obstructions nearer than 100-200m downwind should also be avoided.

iv.   Wind measurements and wind records

Sizing of a windmill for a particular pumping duty is commonly done, where windmills are already in general use, on the basis of the experience of other nearby users of windpower. If a nearby site looks better or worse in terms of exposure to the wind, then some suitable allowances must be estimated to compensate, without too much risk of serious misjudgement. Usually in such situations where windpumps are reasonably common, the suppliers will be able to recommend a suitable size to suit the duty requirement on the basis of past experience.

It is less easy to pioneer the use of windmills in any particular area; in such situations it is necessary to obtain some estimate of the local wind regime. One way to do this is by obtaining data from the nearest meteorological station or airport and making allowances on the basis of the relative exposure of the proposed site compared with the measuring station. Usually the protocol for obtaining this data is to request it from the head office of the Meteorological Department or the Civil Aviation Authority. Unfortunately, however, most small rural meteorological stations were not set up primarily to log wind data, and more often than not, they have incorrectly sited anemometers. Often the anemometers are on 2m tall masts and are surrounded by trees or buildings; any readings from such a site are next to useless for wind energy prediction purposes yet, unfortunately, they are often logged, sent to head office and incorporated in the national data-base, where they distort the apparent wind regime so far as its value for wind power is concerned. Therefore, when using data from a local rural meteorological station, it is strongly recommended that a visit should be made to the station to check whether the data were measured in an acceptable manner to make them of value. Data from international airports (or from major meteorological stations) are usually reliable as the anemometers are normally located at the W.M.O. recommended height of 10m and they will be unobstructed. This is especially true at airports where wind behaviour is of considerable interest from the point of view of aircraft safety. Most such stations log wind speed and direction data continuously on either paper charts or on magnetic tape.

The most basic format for national wind statistics is as in Fig. 126 which shows average wind speeds collected over a number of years, for each month, for a selection of meteorological stations in India. The Indian irrigation season typically occurs in the dry period from January to May, so by inspection it is possible to identify places with seemingly adequate wind regimes during this period; i.e. with monthly means preferably exceeding 2.5-3m/s.

Another format commonly used for the presentation of wind data is given in Fig. 127, which shows a so-called "wind rose"; this includes information on the wind direction as well as its strength. The figure in the middle is the percentage of calm, (defined as less than 3 mph) and the "petals" of the rose are aligned with the points of the compass, (N, NNW, NW, WNW, W, etc.) and shows both the percentage of time the wind blows from each direction; (the concentric circles are 5% intervals) and the mean windspeeds (given in this case in "bins" of 3-8, 9-15, 16-38 and over 39mph. In the example shown the wind is clearly predominantly northeasterly and southwesterly. Wind roses are of most interest for comparing different locations. They are difficult to analyse from the point of view of predicting wind energy availability; raw wind data or even monthly mean wind data are more useful.

If wind records are not available from a sufficiently close or representative existing meteorological station, then it is necessary to set up an anemometer and log wind records for at least one year and preferably two to three years. Obviously this is not a recommendation so much for the small farmer wanting one small windpump, but rather for institutional users contemplating a large investment in wind power which needs to be soundly based on objective wind records. Ideally, about three years of records are required to obtain reasonably representative averages, as mean monthly wind speeds can vary by 10-20% or so from one year to the next. The need for this is of course greater in areas which are thought to be "marginal" for the use of windpumps; it will be shown in the next section that commercial windpumps begin to become economically competitive with engines or other sources of power for water lifting in windspeeds above about 2.5m/s. Therefore, in places which are decidedly windy; i.e. with mean wind speeds almost certainly in excess of say 4m/s, there is no great risk in guessing the mean wind speed when ordering a windpump, and it probably is not worth the trouble and cost of carrying out long term wind measurements in advance. It is possible to "fine tune" the original guess, if it turns out not to be sufficiently accurate, simply by changing the stroke or the pump size. Subsequent windmills can be ordered on the basis of experience with the first one. However, if it turns out that there is just not enough wind, no amount of tampering with the stroke or pump sizes will adequately correct that misjudgement.

The most simple method of measuring mean wind speeds is to install a cup-counter anemometer which simply totalizes the kilometres (or miles) of wind run just as a car odometer totalizes kilometres of road run. By noting the time when each reading is taken, and dividing the difference between two readings by the time interval, it is possible to determine the mean wind speeds over the time period. A mechanical cup-counter anemometer of meteorological office quality costs around US $300 (without a mast - the mast can be improvised with 2" water pipe and guy wires). Ideally such instruments should be read three times per 24 hours, in the early morning, at mid-day and in the evening to allow the diurnal pattern of wind to be recorded. This allows the mean wind speeds for the mornings, afternoons and night periods to be separated. Failing this, an early morning and an evening reading should be taken each day to allow day and night averages to be calculated. Once-a-day or once-a-week readings, providing they are consistently and accurately logged, are however a lot better than nothing, although they will not show diurnal patterns at all.

The effort involved in analyzing raw, continuously recorded data is formidable, so with the recent upsurge of interest in windpower, numerous electronic data loggers have come onto the market which can record wind data in a form that is convenient for "wind energy prospectors". A commonly used approach is to log the frequency with which the wind speed is measured to be blowing within a series of pre-defined speed "bins", such as 0-5 km/h, 5-10 km/h, 10-15 km/h, and so on. If more accurate results are wanted, then narrower bins may be defined to improve the resolution, but this of course requires a more sophisticated logger or more analytical work afterwards.

The most useful starting point for any sophisticated attempt to predict the performance of a windmill in a given wind regime is to create or obtain a velocity-frequency histogram which shows the percentage of the time that the wind blows at different speeds; (as in Fig. 128). This has been constructed from hourly wind data by adding up how many hours in the year, on average, the wind was recorded as having been blowing at a velocity within each pre-defined "bin"; for example, in Fig. 128, the bins are at 1 mph intervals, and there were about 5 hourly records of 0 mph, 100 hourly records of 1 mph, and so on. Clearly an electronic data-logger that automatically measures and records the frequency of wind speeds in predetermined bins makes this task much easier.

It is also quite common to present wind data as a velocity-frequency curve. These are in effect fine resolution velocity-frequency histograms. The wind regime of a given site is characterized by the velocity-frequency curve which will have a similar shape every year and will not vary very much from one year to the next. Velocity frequency curves can be synthesized by a sophisticated mathematical process using what is known as a Weibull Probability Distribution Function, which, providing certain parameters are correctly selected, will produce a passable correlation with natural empirically measured wind regime curves; the analysis required is beyond the scope of this book and is dealt with in ref. [45] among others.

Therefore, the best information that is ever usually available will be an hour-by-hour wind frequency distribution curve for the site. Ideally, this should be combined with data giving the monthly mean wind speeds and the percentage of calm per month. For irrigation pumping it is of critical importance to consider the wind regime during the month of maximum water demand; annual averages are not good enough for this.

v.   Windpump manufacturers' performance claims

The easiest method of estimating the performance to be expected from a windpump is to use manufacturers' data as printed in their brochures. For example, Fig. 129 shows a table and performance curves for the "Kijito" range of windpumps, based on the "IT Windpump" and made in Kenya, (see Fig. 116). The table indicates the average daily output to be expected at different pumping heads for the four sizes of Kijitos in three different average speeds, defined as "light" 2-3m/s, "medium" 3-4m/s and "strong" 4-5m/s while the curves reproduce these results just for the "medium" wind speeds. It is interesting to note how sensitive windpumps are to windspeed; the smallest machine with a 12ft (3.7m) rotor will perform in a 5m/s wind almost as well as the largest machine (24ft or 7.3m) does in a 3m/s wind; this is because there is 4.6 times as much energy per unit cross-section of a 5m/s wind as in a 3m/s wind as a result of the cube law.

A problem with manufacturers' performance claims is that some brochures include inaccurate, unreliable or even incomplete data. For example, it sometimes is not clear what wind speed applies for the manufacturer's claimed outputs. There is a tendency for manufacturers to quote performance figures for unusually high average wind speeds, no doubt because this makes the performance look more impressive, and they then give rules of thumb which in some cases do not seem accurate for reducing the outputs to more realistic levels for more common wind speeds.

The difficulty inherent in monitoring the performance of a windpump often prevents users from actually checking whether they are getting what they were promised; if a cup counter anemometer and a water meter is available, it is possible to measure the wind run and the water output over fixed periods of time (either 10 minute intervals or daily intervals - short-term and long-term

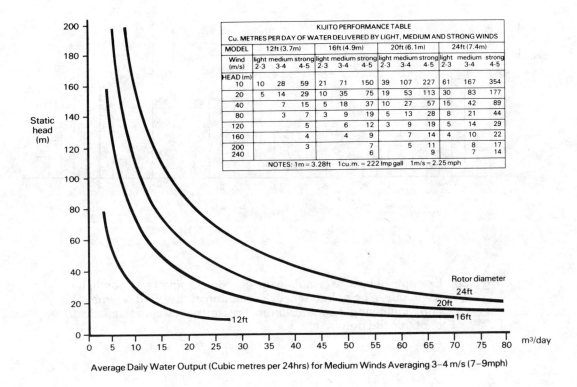

Fig. 129 Manufacturers' performance data for the Kenyan-made 'Kijito' windpump range based on the I.T. windpump (see also Fig. 116)

Table 19 CALCULATION OF WINDPUMP OUTPUT USING "BINNED" WINDSPEED DATA COMBINED WITH PERFORMANCE DATA

		Annual output of water for a given wind régime		
Wind speed		Annual duration	Output rate	total output
m.p.h.	m/sec	hours	m³/hr	m³
7	3.15	600	0.3	180
8	3.6	500	1.4	700
9	4.05	500	2.3	1,150
10	4.5	400	3	1,200
11	4.95	500	3.7	1,850
12	5.40	450	4.2	1,890
13	5.85	450	4.7	2,115
14	6.30	300	5.2	1,560
15	6.75	300	5.7	1,710
15 plus		1,700	6	10,200
		Total 5,700		Annual Total 22,555 m³

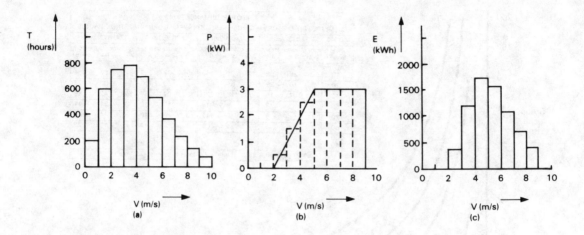

Fig. 130   Example of how to calculate the energy output of a windmill by using the velocity frequency distribution of the wind regime (a) and multiplying it by the windmill performance characteristic (b) to obtain the output (c)

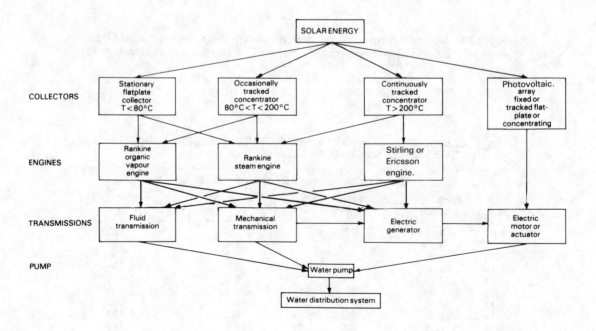

Fig. 131   Feasible options for solar-powered pumping systems

- is recommended). The mean wind speed and mean water output over these periods can be logged and then plotted as a scattergram. When enough points are obtained on the scattergram, a "best-fit" curve can be drawn to obtain the performance characteristic. It is recommended that institutional users and others who need to procure numerous windpumps should seek performance guarantees from their supplier and should make some attempt to verify whether the performance is being achieved, for example, in the manner just suggested. In some cases sub-optimal performance can occur simply because the wrong pump size or wrong stroke has been used, and considerable improvements in performance may result from exchanging the pump for the correct size or from altering the stroke.

4.7.4 Windpump Performance Estimation

i. General principles

To size a windpump for irrigation purposes will usually require an estimate to be made of the week by week or month by month average output. One method for making such an estimate is to combine data on the known performance of the windpump at various hourly average wind speeds with data from a wind velocity distribution histogram (or numerical information on the number of hours in the month that the wind blows within pre-defined speed "bins"). This is illustrated by Table 19, which gives the expected output of a windpump in various windspeeds, and the statistical average number of hours that the wind blows within each speed range, (or speed "bin" is the favoured jargon). Hence, the total output for each speed bin is obtained by multiplying the output per hour at that speed and the number of hours at which that speed is likely to recur. By adding together the output for each speed bin we arrive at the total annual output. The importance of doing this monthly is that quite often the least windy month will have a mean windspeed of only around 60 to 70% of the annual mean wind speed, so the available wind energy in the least windy month can be as little as 20% of what can be expected for a mean windspeed equal to the annual average wind speed. Therefore if annual averages are used, a considerable margin of safety is necessary to allow for "least windy month" conditions, (assuming irrigation water is needed in the least windy month or in a month with a mean windspeed below the annual average.

A graphic way of achieving the same result as used in the Table 19 is illustrated in Fig. 130. Here the wind velocity frequency histogram, (preferably for a month at a time) shown in (a) is multiplied by the windpump performance characteristic shown in idealised form in Fig. 130 (b); in this example a windpump is assumed which starts at a nominal 2m/s, produces 3kW of output at its rated speed of 5m/s and is fully furled at 9m/s. A windpump performance histogram using similar speed "bins" to diagram (a) is constructed over the windpump performance characteristic. Finally the windpump's generalised performance histogram in (b) is multiplied by the wind velocity distribution histogram in (a) to arrive at a performance prediction histogram (c). To illustrate the mechanics of multiplying two histograms together, the first speed bin in (a) is 200h, the first in (b) is zero kW, so their product in (c) is zero kWh; the second is 600 x 0 = 0, the third is 750hours x 0.5kW = 375kWh, the fourth 800h x 1.5kW = 1200kWh, and so on.

ii. Simple "rule-of-thumb" approach

The problem with the above methods is that the result is only as good as the wind and performance data used, which all too often are unreliable. Therefore, if only approximate data are available, a simpler rule of thumb may be more appropriately used than attempting any detailed analysis, (and is often useful anyway as a cross check on other methods). This method was originally proposed in [45] and [51], and provides a reasonable method of estimation.

The rule of thumb assumes that a windpump system will, on average, be 17% efficient in converting wind energy into hydraulic output, which in many cases is probably not a bad estimate (the losses in a windpump system and the total efficiency that can be expected are discussed in more detail in the next section of this chapter). The average hydraulic output power for a windpump of about 17% average efficiency will be:

$$P = .17 \times \tfrac{1}{2}\rho V^3 \text{ W/m}^2 \text{ of rotor area}$$

because the density of air at sea level is approximately 1.2kg/m³, it follows that:

$$P = .17 \times \tfrac{1}{2} \times 1.2 \times V^3 \text{ W/m}^2$$
$$\text{so } P = .1V^3 \text{ W/m}^2$$
$$\text{or } P = .1 \times \tfrac{1}{4}\pi D^2 \text{ V}^3 \text{ watts for a rotor diameter D metres}$$

If $.1V^3$ is multiplied by the time interval in hours applying to the average windspeed used, then the output can be calculated. For example, if $\overline{V}$ is the daily mean windspeed (based on 24 hours), then the daily hydraulic energy output will be:

$$E = 24 \times .1\overline{V}^3 \text{ Wh/m}^2 \text{ of rotor area}$$

if this is multiplied by the rotor area in square metres, it gives the daily hydraulic energy output. Dividing the number of watt-hours per day by 2.725 converts this to a daily "cubic metres-metre" product, or m³.m (i.e. output in cubic metres times head in metres). Although an unusual method of expressing energy, this can readily be converted to a daily output of water at any particular pumping head; eg.:

$$1{,}000 \text{ Wh/day (or 1 kWh/d)} \text{ (hyd)} = \frac{1{,}000}{2.725} = 367 \text{ m}^3.\text{m}$$

which is 36.7m³/day if lifted through 10m, or 3.67m³/day through 100m, etc.

iii. Overall system efficiency

Table 20 indicates the efficiency factors relating to windpumps, and shows that between 7 and 27% of the value obtained for the energy in the wind (using the mean wind speed) can be converted to hydraulic energy. Because the speed of the wind continuously varies, the actual energy in the wind is considerably greater than if the wind blew continuously at the mean wind speed. The exact proportion by which the energy available exceeds the figure that can be calculated based on the mean depends on the shape of the wind velocity distribution curve, but typically it can be twice the figure arrived at using the mean. This explains why there is an "efficiency" factor of 180-250% used to allow for the underestimate which would otherwise occur if the mean wind speed were simply cubed with no other allowance.

Table 20   FACTORS AFFECTING WINDPUMP SYSTEM EFFICIENCY

factor	typical efficiency
rotor to shaft	25 - 30
shaft to pump	92 - 97
pump	60 - 75
wind regime % energy capture	30 - 50
actual wind energy to wind energy calculated using mean wind speed	180 - 250
TOTAL EFFICIENCY OF ENERGY CONVERSION BASED ON MEAN WIND SPEED	7 - 27 %

This result indicates that the figure of 17% overall efficiency used to produce the $0.1\overline{V}^3$ rule of thumb described above may vary with different windpumps and wind regimes by plus or minus 10% of the total energy available. Therefore, the rule of thumb coefficient used of 0.1, and hence the predicted output using this method, may generally be expected to be within the range of 0.6 to 1.6 times the result obtained by using 0.1 and the mean wind speed.

iv..   Wind requirements for economic operation

A convenient method for costing and comparing windmills is to estimate the total installed cost of a system as a function of rotor area (e.g. in dollars per square metre of rotor). Since the energy output is a function of the wind regime combined with the system efficiency and its rotor area, the unit cost of the rotor area will give an indication of the cost of energy in a given wind regime if a uniform system efficiency is assumed, (or upper and lower limits such as indicated in the previous section may be used).

Wind pump prices correlate quite closely with the area of rotor being purchased. Prices for industrially manufactured windpumps, (not including shipping and installation) are in the US $200-400/m² range, while wind generators tend to be about three times as expensive per unit of rotor area (for small machines) or twice as expensive for larger machines. Windpumps made in developing countries can be significantly cheaper; for example the "Tawana", Pakistan-made version of the I T Windpump costs about $130/m² (in

1985). "Do-it-yourself" windpumps, built in the village, such as are used in Thailand are very much cheaper still, but the cost depends very much on what assumptions are made on the value of the construction labour.

Combining these costs with the previously given performance assumptions will show that at current prices, most windpumps need mean winds speeds in the region of 2.5 to 3m/s to begin to be economically attractive, and wind generators need 3.5 to 5m/s. Moreover the cube law ensures that the economics of windmills improve dramatically at higher windspeeds, making them a most economically attractive technology in most wind regimes having mean windspeeds exceeding say 5m/s.

4.8 SOLAR POWER

Small-scale irrigation pumping is one of the more attractive applications for solar power. Solar radiation tends to be at its most intense when the need for pumped water is greatest and the energy supply is available at the point of use, making the farmer independent of fuel supplies or electrical transmission lines. The main barriers at present to the wider use of solar pumps are their high costs combined with general unfamiliarity of the technology. A cheap and reliable solar pump, which could quite possibly be developed within the next few years, will have the potential to revolutionize Third World agriculture. This is why it is worth being aware of the potential of this promising new technology, even though it is probably not yet economically viable for irrigation.

Many solar pumps have been built and operated, so their technical feasibility is proven, but the technology is still immature with production running in dozens or hundreds of units per year rather than the thousands that must be manufactured before costs drop. Also, solar pumps tend to become economically viable in water-supply applications sooner than for irrigation, due to the much higher value that can be placed on drinking water; in fact an economic case can already be made for using solar pumps for village water supplies given favourable operating conditions. Solar pumps for irrigation are only currently economically viable at very low heads where the power demand is extremely small. Nevertheless, significant technical developments coupled with cost reductions are being achieved and it can be expected that reliable and economically viable solar irrigation pumps will be available within the medium to long term.

4.8.1 Background and State-of-the-Art

There are two main methods for converting solar energy into power for driving a pump. Solar thermodynamic systems depend on using the heat of the sun to power an engine (steam or stirling cycle), while solar photovoltaic systems convert solar radiation directly to electricity by means of photocells, and hence they can power an electric pump. The most feasible routes for applying solar energy to pumping water are shown in Fig. 131.

Paradoxically, solar thermodynamic experiments date back over a century, yet today such systems remain immature compared with the much more recent solar photovoltaic alternative based on technology discovered as recently as the 1950s.

The first successful solar thermodynamic systems were developed in France during the mid-nineteenth century; see Butti and Perlin, [52] and Daniels, [53]. By 1900 most of the development effort had shifted to the USA, where several people were seeking to develop commercially viable solar pumping systems. However, although several solar steam engines were successfully demonstrated, they tended to cost several times as much as a conventional steam engine of the same power (although their fuel costs were of course free), which limited their commercial success. This work culminated in the construction in 1914, by an American called Frank Shuman, of what even today remains one of the largest and, for that matter most technically successful solar thermodynamic pumping systems. Developed in the USA, but installed at Meadi in Egypt to pump irrigation water, this unit incorporated hot water storage and could therefore drive an irrigation pump for 24 hours per day. After a few teething troubles, the Meadi solar pump was shown to develop 55 hp (40kW) and it could pump up to 1,300 m³/h (360 litre/sec). Under Egyptian conditions at that time the plant could pay back in competition with a steam engine in 2 years and could completely pay for itself in 4 years. Enormous interest ensued, and ambitious plans to install similar solar pumps in other parts of the world were initiated, but the First World War broke out and Shuman, the driving force behind solar pumps at that time, died before it ended. The cheap oil era that followed led to a lack of interest in solar power for pumping until the oil price increases of the 1970s.

Numerous researchers initiated laboratory research work on solar thermodynamic systems during the 1970s, and SOFRETES, a French company, began manufacturing a low temperature solar pumping system, nominally capable of about 500W-1000W of pumped output, which was distributed with French government support to a number of developing countries in the late 1970s on a pilot demonstration basis. Also, a number of quite large solar thermal installations, some for pumping water and some for electricity generation, were completed in the USA and in Africa. Unfortunately, the majority of these latterday solar thermodynamic systems performed poorly and many also proved troublesome to operate reliably under field conditions. Today the French company that produced the solar thermodynamic systems is no longer in business and no other manufacturer has a proven solar thermodynamic system commercially available.

In 1979, the UNDP (United Nations Development Programme) initiated a Global Project specifically to evaluate small scale solar pumps for irrigation, with the World Bank as the executing agency; (project UNDP GLO/78/004 followed by GLO/80/003). The UNDP Project field trials were held in Mali, Sudan and the Philippines, and it was necessary to select solar pumps which were at least nominally commercially available and which showed reasonable prospects of being able to function adequately in the field. when bids were invited from suppliers all over the world, it transpired that only one solar thermodynamic solar pump was available for procurement by the project in 1980, and this only barely fulfilled the minimum requirements to justify field testing it. In contrast, no less than eleven solar photovoltaic pumps were short-listed from a still broader choice. As these projects proceeded the emphasis moved from the use of solar pumps for irrigation to their use for water supply duties, where they are more immediately economically viable. Numerous reports were produced for this project; eg. [54],[55] and [56], plus a "Handbook on Solar Water Pumping", [57], which is intended to help those seeking more detailed information on selecting and sizing solar pumps.

Therefore, solar thermodynamic small scale systems have not really succeeded, despite a century of development with early signs of promise, but the much newer photovoltaic technology, although still expensive, works reasonably successfully. Photovoltaic systems use photo-electric cells, which can convert sunlight directly to electricity. This technology had its origins at the Bell Laboratories in the USA in the early 1950s. Solar photovoltaic cells were originally developed to power space satellites, which in common with many rural parts of the earth's surface, need a small independent power supply. This technology therefore started at the frontiers of science on an "expense no object" basis. Therefore, the earliest terrestrial solar power cells, which became available in the 1960s, were prohibitively expensive for such applications as irrigation pumping.

The considerable cost of solar photovoltaic cells, which are generally made from a highly purified form of silicon (an abundant element) relates mainly to the high overheads associated with the investment in sophisticated manufacturing facilities needed, rather then to the cost of raw materials. There are therefore major economies of scale relating to volume of production, so costs have consequently been declining steadily over the last few years. By 1984, solar modules (the basic "building brick" for a solar array) could be purchased on the world market for US $8.00 per rated peak watt. Solar photovoltaic systems have their power capability defined in "peak watts" which is the electrical output that would be produced in direct sunshine of $1,000W/m^2$ strength when the solar module has a cell temperature of 28°C. Many predictions of radical reductions in solar cell costs "by the mid 1980s" were made in the buoyant period for renewable energy development of the late 1970s; in the event these have not yet happened, but nevertheless, significant and useful cost reductions have been achieved (50% in real terms between 1980 and 1985). It can reasonably be expected that the present technology solar cells will come down in price in real terms by another 50% as production volume increases, and this will be enough to make them worthy of serious consideration for general small scale irrigation pumping at heads up to about 6m. Possible new solar cell technologies could conceivably result in even greater cost reductions in five to ten years' time, which would make this technology decidedly attractive compared with the other options, (most of which are getting more expensive in real terms).

4.8.2 Principles of Solar Energy Conversion

Because solar thermodynamic pumping systems, although experimented with for over a century, remain generally immature, it is proposed to deal only briefly with them and to concentrate on the currently more promising photovoltaic technology.

i. Thermodynamic systems

Thermodynamic systems can be further sub-divided into three main categories, operating at low, medium and higher temperatures with a "Rankine" or steam engine at the lower or medium temperatures and a stirling engine at the higher temperatures.

As indicated in Fig. 131, thermodynamic pumping systems always include a solar collector, which collects solar heat and transfers it to a working

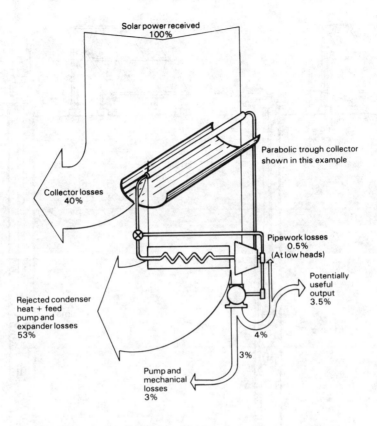

Fig. 132   Losses in a typical solar thermodynamic pumping system

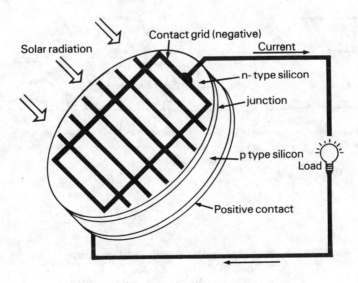

Fig. 133   Construction of a silicon photovoltaic cell

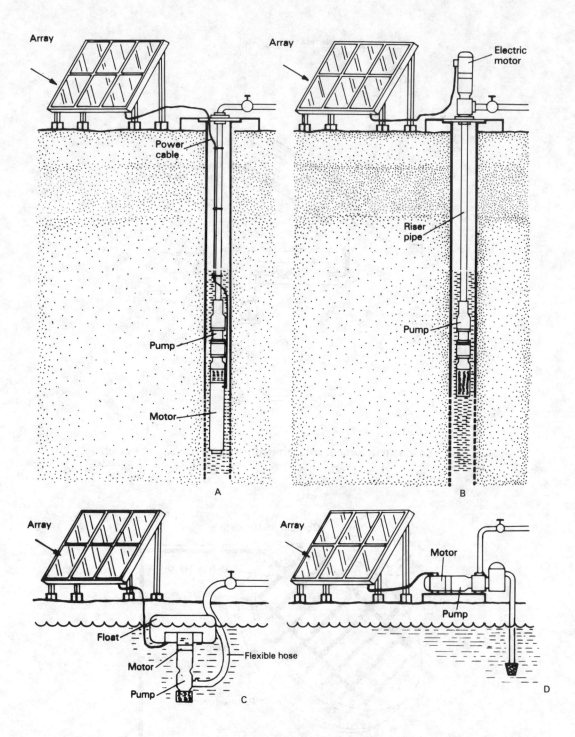

Fig. 134 Examples of solar pump configuration
A. submerged motor/pump set
B. submerged pump with surface motor
C. floating motor/pump set
D. surface motor with surface mounted pump

fluid; an engine, which takes heat from the working fluid and converts this to shaft power and a transmission system to connect the engine to a pump.

Low temperature thermodynamic systems use what are called "flat plate collectors", which usually consist of a flat absorber panel containing passages through which the working fluid circulates. These are generally painted matt black and mounted in a shallow insulated housing behind a glass window of similar area to the panel. The absorber panel is mounted at an angle to the ground of approximately the latitude of the location, and facing towards the equator. This has the effect of maximizing the amount of solar energy received by the absorber.

The glass window readily admits light, but inhibits the outflow of the heat which is produced; this is commonly known as "the greenhouse effect". This effect can be enhanced with double glazing, but this also increases the costs, since the glass is one of the more expensive components. Similarly, the absorber panel can be coated with a special "selective surface" which also enhances the efficiency of absorption of solar radiation, but at increased cost compared with regular black paint. Flat plate collectors of this kind can only achieve working temperatures of up to about 80°C, but more usually will operate at about 60°C. Flat plate collectors are therefore used with low-boiling point working fluids such as ammonia, the freons (fluoro-hydrocarbon fluids which are used in air conditioning and refrigeration circuits) or even butane (the latter poses some fire hazard if it leaks). However, even with low boiling point fluids, the efficiency of a rankine engine will be very low, because the thermodynamic efficiency is, for fundamental physical reasons, a function of the temperature difference between the hot vapour admitted to the engine and the cool vapour that is exhausted. The exhausted vapour can only be cooled to the condenser temperature which will generally be the temperature of the pumped water which is usually used to cool the condenser. Therefore it is only possible to achieve a temperature difference (sometimes known as "delta T") of about 30°C with flat plate collectors.

If higher temperatures are required it becomes necessary to focus, or concentrate, the sunlight on a smaller area than it would naturally fall on. This is normally achieved by using a parabolic mirror. The simplest arrangement is a line focussing parabolic trough collector, which focuses the light falling on it into a line, as in Fig. 132; greater temperatures and concentration factors can be achieved with a "point focus" parabolic dish concentrator. These offer improved thermodynamic efficiency in terms of a greater "delta T" (line focussing achieves 100-200°C while point focussing achieves 200-500°C). Unfortunately focussing, concentrating solar collectors have two major disadvantages (which apply equally to the use of focussing with photovoltaics as well as with thermodynamic systems):

a. they need to be aimed at the sun so that the sun's rays are focussed on the heat absorber; the greater the concentration, the greater the accuracy with which the device must be aligned with the sun. To do this requires a mechanical means to make the collector track the sun;

b. they can only focus direct rays (beam radiation) and therefore cannot make use of scattered light (diffuse radiation) which is available to flat plate collectors. In areas with haze, atmospheric dust, high humidity or scattered clouds there is much diffuse radiation, often amounting to as much as 30% or more of the total solar irradiation, which is not accessible to a concentrating collector but which can be used by a flat plate collector.

Therefore, although a concentrator improves the thermodynamic efficiency of a solar pumping system, and thereby reduces the area of solar collector needed to achieve a given power output, the added complication increases the cost per square meter of solar collector and also reduces the available solar energy by limiting usage to direct beam radiation. Analysis by Halcrows and I T Power [54] [55], indicates that a compromise of using low concentrating factor line focussing collectors appears to be the most cost-effective approach with current solar collector costs, but that future technical developments could make point-focus systems competitive if a clever, low cost and reliable parabolic dish and sun tracking mechanism is developed. But this would only apply in locations (such as deserts) with a high proportion of direct beam radiation.

Fig. 132 is a schematic diagram of a solar thermal pumping system, showing the energy flow and the principal losses. A good solar collector will typically collect 60% of the solar energy falling on it. Of this 60%, about 7% will be converted to shaft power and the rest is lost as rejected condenser heat, feed pump and expander losses, and finally perhaps 50% of what is left, i.e. 3.5% of the original input, will be converted into useful hydraulic pumped power. This is an example of the better types of small solar thermal systems, many that have been developed are not nearly as good as this in practice.

ii. Photovoltaic power systems

Systems of this kind depend on a property of certain semi-conductors to generate electricity when exposed to light. The physical principles are explained in broad detail in [57] or in depth in [58]. A number of different photovoltaic cell materials exist, but the only commercially proven power cells currently available are based on thin wafers cut from purified crystals of silicon. The majority of these are sliced from single crystals of silicon and are known as "mono-crystalline" cells. However, a recent process involves the growth of several crystals simultaneously, which when sliced yield so-called "poly-crystalline" cells. Both technologies are competitive at present.

Silicon is an abundant element, but the purification, crystalization and slicing process requires high technology and is expensive. Work is in hand on a variety of cell technologies that promise in some cases improved efficiency and in others lower costs, but none are as yet at the production stage. Some of the so-called "thin film" technologies under development promise substantial reductions in costs which could greatly enhance the scope for using photovoltaics when they become commercially available.

Normal practice with mono- or poly-crystalline silicon cells is to slice them into wafers, which may be circular or square and are typically 100mm across, but quite thin (about 1mm or less). The front and rear surfaces are doped with impurities to create the necessary semi-conductor properties. The rear surface is metalised, while the front has a fine pattern of conducting metal collectors plated on to it which are usually connected by two thicker conductors; see Fig. 133. Light falling on the cell produces a potential difference (or voltage) between the conductors on the top surface and the metalised rear surface. Each cell can develop about 0.4 volts when under load, (0.6V on open circuit), so by connecting a string of cells together in series, a higher voltage can be created.

Solar cells are delicate, and expensive, and need to be safely enclosed behind an appropriate window to let the light in. Therefore strings of cells are generally mounted in glass fronted frames known as "modules". Toughened glass is usually used for the cover (this is sometimes given a matt finish to minimise reflection) and the cells are encapsulated behind the glass in plastic with either a metal or glass rear surface. The frame is commonly made of light-alloy and a good seal is needed between the frame and the glass.

Typically, cells are mounted in strings of 36 per module, giving a nominal voltage of 14-16V. 100mm diameter cells typically produce about 1W of electrical output when exposed normal to full strength sunlight amounting to 1000W/m², so a standard module of 36 x 100mm cells will usually be rated at 35W (recent types with larger and/or more efficient cells are rated at 40W and more).

Solar modules are the most expensive part of a solar photovoltaic system, costing in 1985 typically US $7.00/W(p) (per peak watt); i.e. a standard 40W module currently costs around $280. Fortunately, good quality modules are highly reliable and should last 20 years or more, providing they are not subject to physical damage to the glass covers. A serious worry, once their high intrinsic value becomes more widely known and a "second-hand market" develops, is the possibility of theft.

Modules are mounted on a supporting structure in a frame and this assembly is known as a photovoltaic array which can simply be connected by electric cable to the motor/pump sub-system. Examples of four typical types of solar photovoltaic pumping systems are shown in Fig. 134.

In a few cases arrays have been developed which track to follow the sun; this of course increases the amount of sunlight intercepted as the array will always face the direct beam radiation. However, the complication and expense of providing mechanical tracking has not generally been found worthwhile so it is not common practise. A few portable, or semi-portable small scale solar pumping systems exist in which the arrays can be manually orientated by moving them bodily; here some significant advantage at little effort or cost is gained by facing the array(s) south east in the morning and south west in the afternoon when in the northern hemisphere (or NE and NW respectively in the southern hemisphere).

Most arrays are generally designed to carry the modules at a fixed tilt angle which maximises the amount of sunlight received over the year. It so happens that the theoretical angle for this to be achieved coincides with the angle of latitude of the location. This is because that is the angle which will place the array perpendicular to the sun at solar noon on the equinoxes, and will therefore make the divergence of the sun's rays from perpendicular to the plain of the array a minimum at all other times. In practice, the actual angle for the array is sometimes varied so as to bias the optimization to suit a season having more cloud cover; for example, in areas with a marked rainy season it may be advantageous to incline the array more normal to the sun in that season, and sacrifice a little solar energy in the dry season when perhaps more energy than necessary can be intercepted. Also, although mounting the array at zero angle (i.e. horizontal) is the optimum angle for the equator, it is normally recommended that arrays be mounted at angles of at least 10° so that there will be good rainwater run-off, which helps to keep them clean. It is often advantageous to have an array with a capability of having its angle adjusted manually every few months, as this can improve the output by 10% or more, over the year.

The performance characteristics of silicon solar cells are shown in Fig. 135. It can be seen that the efficiency of energy conversion is in the region of 10%; e.g. in peak sunlight (which yields approximately 1000 W/m²) 1m² of silicon cell surface will yield about 100W of electrical output. Unfortunately, the efficiency of solar cells declines as they get hotter as shown in Fig. 135 A. Most solar cells attain temperatures in the range 50-60°C in full sunlight conditions, but their power rating is generally quoted by manufacturers for an operating temperature of 25°C, which coincides with laboratory conditions for a standard method of testing. They therefore will never achieve their rated output in practise under normal sunshine, but sizing methodologies, such as are described later, take account of this discrepency.

Fig. 135 B indicates the voltage-current (V-I) characteristic of a solar cell, with 1000 W/m² irradiance. At the one extreme you can short-circuit the cell and get a maximum current Isc (the short-circuit current) of about 30mA/cm² of cell, but the voltage will be zero. At the other extreme, on open circuit, there is zero current, but an open-circuit voltage, Voc of about 0.55-0.60V per cell, (regardless of its size). The maximum power occurs when the load applied is such that the maximum value of the product of V and I is obtained (electrical power in watts is numerically volts times amps). This occurs near the "knee" of the V-I curve at around 0.4V per cell, as indicated by the dotted curve showing power. As the efficiency is exactly proportional to the power output, the efficiency curve is identical in shape to the power curve.

The sun mostly shines with a strength of significantly less than 1000W/m², (which is a mid-day intensity), so Fig. 135 C indicates how, if the irradiance is reduced by say 80% to 200W/m², the current density declines proportionately, but the voltage stays the same. The power from a cell is therefore almost linearly proportional to the intensity of illumination from the sun. Because a voltage is produced even in low levels of irradiance, solar photovoltaic systems can function in low levels of sunlight, depending on the nature of the electrical load and what the threshold power for starting happens to be.

A few manufacturers have produced solar photovoltaic systems with concentrating collectors, using mirrors or lenses to intensify the strength of sunlight on the cells. This allows a smaller area of cells to be used, but the added expense of mirrors or lenses seems to negate any saving on cells, especially bearing in mind the other disadvantages of concentrating collectors discussed earlier.

iii.   Photovoltaic motor-pump sub-systems

Photovoltaic pumping systems require, in addition to the array, a so-called "sub-system" which consists of at least an electric motor powering a

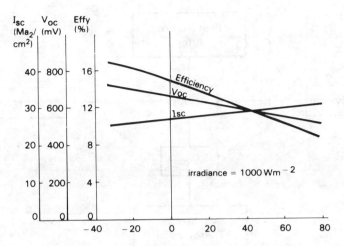

A — Dependence of Efficiency, I_{SC} and V_{OC} on Cell Temperature

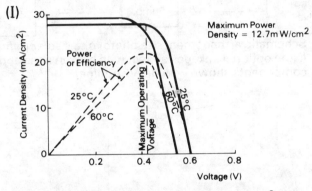

B — Effect of Cell Temperature on I V Characteristic (for 1000 W/m²)

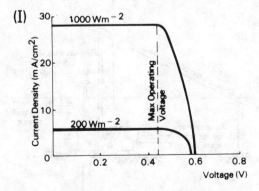

C — Effect of Change in Irradiance on I V Characteristic

Fig. 135 Performance characterists of silicon photovoltaic cells

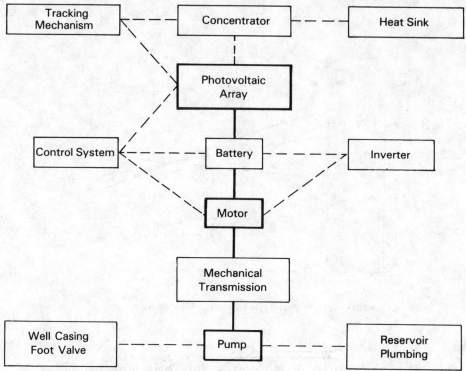

Fig. 136 Schematic arrangement of a photovoltaic solar pumping system. (Note optional linkages shown with broken lines and essential components shown with heavy outline)

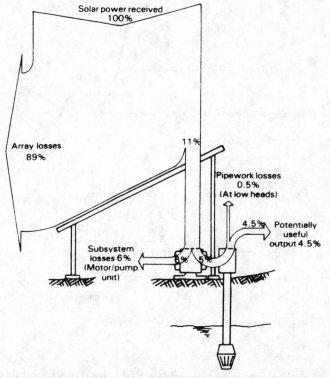

Fig. 137 Losses in a typical solar photovoltaic pumping system

pump. Fig. 134 illustrates various commonly used configurations that can comprise a photovoltaic pumping sytem. Fig. 136 shoes schematically the necessary (and optional) components of a solar photovoltaic pumping system and how they interact.

Because the output from an array is DC (direct current), either a DC electric motor needs to be used, or alternatively an inverter can be used to synthesize AC so as to allow a conventional mains power AC electric motor to be used. The disadvantage of using an inverter is its cost and the power loss inherent in its use, but the advantage is that many relatively low cost, mass-produced, standard electric pumps can be used.

Where DC motors are used, the most efficient types are permanent magnet motors, where the flux is provided by magnets rather than an electro-magnetic field coil. A problem with DC motors is that carbon brushes are generally needed and these are a potential source of trouble as they wear and need regular replacement. Brushless, maintenance-free, DC motors have recently been developed however, which have electronic circuitry to perform the same function as a commutator and brushes.

Much of the discussion in Chapter 4.6 in the context of mains (or generating set) electricity applies to the motors and pumps of photovoltaic electric systems. For example, combined submersible motor pump units may be used, (either the borehole type or surface water type) which need either AC or brushless DC motors since they are sealed and therefore not suited to brushed motors. Surface suction pumps direct coupled to motors may be used, or surface mounted motors driving submerged pumps via a long shaft. Also, a transmission may be used to convert the rotary output of a motor to a slower reciprocating output and to drive a standard piston pump.

For irrigation duties, generally at low heads, the most common and useful system is either a submersible motor pump unit, usually suspended from a float for pumping surface water, or a fixed installation usually using a surface mounted motor driving a centrifugal or multi-stage turbine pump via a long shaft. Surface-suction centrifugals also have a role, but self-priming capabilities are essential with a solar power source or the user may need to reprime the pump everytime a cloud passes by.

Because the motor-pump combination will have an optimum performance at a certain voltage and current, the photovoltaic system designer tries to match a motor-pump unit to the array so that it will electrically load the array under all typical irradiance levels so as to keep the voltage and current levels as near to the "knee" of the photovoltaic module characteristic as possible. There is an optimum characteristic curve for any array where the output is maximised under all solar conditions, so the aim is to provide a motor and pump with a characteristic such that with varying currents caused by changes in irradiance level, the voltage requirement follows the array maximum characteristic curve as closely as possible. With good design, centrifugal pumps can be sized to load a system very close to this optimum right across the operating range. Positive displacement pumps, on the other hand, need varying voltage to change their output and therefore they inherently fail to match well with a solar power system. Another difficulty with positive displacement pumps is that they generally need more torque (and hence current) to start than to keep going, which makes starting in low levels of irradiance problematic unless some artificial method to overcome the problem is used.

A general solution to matching arrays and motor-pump units (and to allow for variations in their load lines when operating off the design head, for example), is to use an electronic power conditioning device called a Maximum Power Point Tracker (MPPT). This can convert a DC input at one voltage to a DC output at another voltage. It is possible to sense the output current and voltage and provide a control function which will automatically adjust the output voltage to obtain maximum power for the irradiance level and pump load applying at any time, generally using a microprocessor to make the logical decisions. This approach results in extra cost for the MPPT, plus a small parasitic power loss inherent in any extra component, but in many cases, particularly with positive displacement pump systems, it greatly improves performance. The gains from an MPPT are less worthwhile in the case of well-matched low head direct-driven centrifugal pumps of the kind which tend to be most often used for irrigation duties.

A controversial technical question is whether batteries should be used with solar pumps. While they are not essential, and therefore not often used, they can provide a useful element of energy storage, allowing the system to work when the user wants to use it rather than when it happens to be sunny. Batteries also provide power conditioning capability as an alternative to a MPPT in that they can deliver a different current to that received. However their life under tropical conditions tends to be limited to four or five years, and they need regular topping up with distilled water unless more expensive sealed batteries are used.

As always, each component causes a loss and an inefficiency; the power flow through a typical photovoltaic pumping system is illustrated in Fig. 137. This indicates that, at best, the following efficiencies can be expected:

$$\begin{array}{ll} \text{photovoltaic modules} & 11\% \\ \text{motor-pump unit, cables and controls} & 45\% \end{array}$$

Therefore the actual hydraulic output, ignoring pipe losses will be about 5% of the power of the sun arriving through the window of the array. Pipe losses, even with a well designed low head system, may be another 10% of the hydraulic output, or 0.5% of the total input, giving at best an overall efficiency of 4.5%.

Tests, such as are reported in ref. [59], have shown that the best systems do achieve this level of efficiency, but many systems are not nearly as good being typically only 2-3% efficient or worse. Efficiency is an important factor in selecting a solar pump, because with the high price of photovoltaic arrays low efficiency implies a need for a proportionately larger and more expensive array. A measure of this can be gained by an example of comparing the use of a 70% efficient motor with an 80% efficient one for a 500W(p) system. At this power level, 10% extra array size to make up for the less efficient motor will cost approximately US $300-400. Since the difference in cost between an efficient motor and an inefficient one is usually much less than this, it is obviously worth buying the best motor available. This decision is of course usually made by the system manufacturer and not the user, but the informed user can at least question the choices made by manufacturers by studying their specifications before buying.

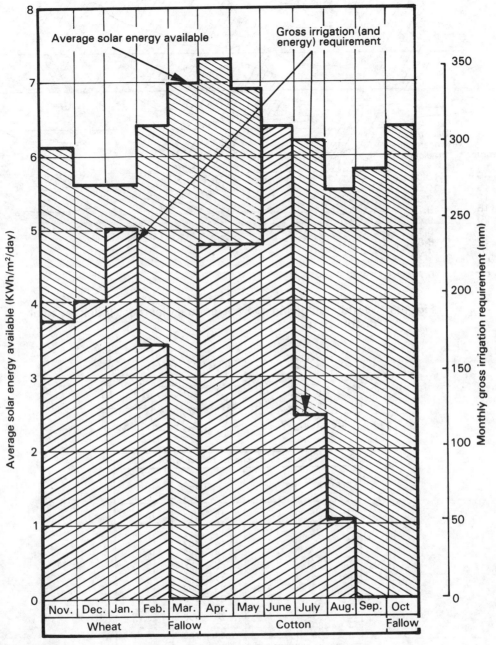

Fig. 138 Solar energy availability compared with crop irrigation water demand (Lake Chad region)

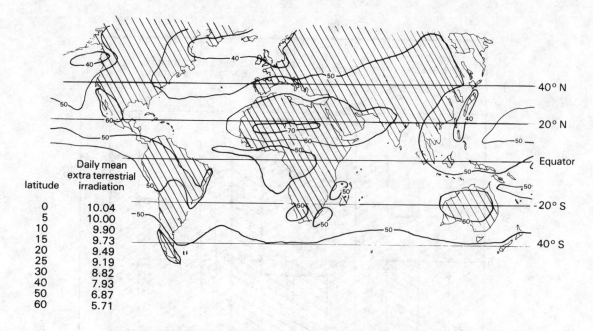

latitude	Daily mean extra terrestrial irradiation
0	10.04
5	10.00
10	9.90
15	9.73
20	9.49
25	9.19
30	8.82
40	7.93
50	6.87
60	5.71

Fig. 139 World map giving average annual Clearness Index for solar distribution

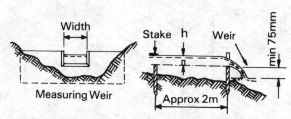

Discharge in litres per second per 100mm width of weir.

DEPTH h in	mm	l/sec	DEPTH h in	mm	l/sec
1	25	0.2	15	381	10.4
2	50	0.5	16	406	11.4
3	76	1.0	17	432	12.4
4	102	1.4	18	457	13.5
5	127	2	19	482	14.7
6	152	2.7	20	508	15.8
7	178	3.4	21	559	17
8	203	4.1	22	584	18.1
9	228	4.9	23	609	19.4
10	254	5.7	24	635	20.6
11	279	6.6	25	660	21.9
12	305	7.5	26	686	23.1
13	330	8.4	27	711	24.4
14	355	9.4	28	736	25.7

In case of small streams or channels a temporary measuring weir can be easily erected as indicated. It may be made of strong timber or metal sheet, with the bottom and sides of the rectangular notch bevelled, to a width of about 2mm. The distance from the bottom of the notch to the downstream water level should be at least 76mm (3 inches) and sufficient to allow for complete aeration. This weir is let into the stream and made watertight with clay or plastic sheet so that all the water will pass through the rectangular notch. For accurate measurements a stake should be driven into the stream bed about 2 metres upstream, the top of the stake being level with the crest of the weir. The depth of water flowing over the weir can be obtained by measuring the height (h) from the water surface to the top of the stake. The flow may then be calculated from the table below.

Fig. 140 Flow measurement with rectangular weir (after BYS Nepal)

4.8.3 The Solar Energy Resource

The average value of solar irradiance just outside the earth's atmosphere is 1,353 W/m². Solar radiation is attenuated as it passes through the atmosphere so that the maximum level normally recorded at sea level is around 1,000W/m². This is made up of two components, the direct beam radiation from the sun and the diffuse radiation consisting of light scattered by the atmosphere from haze, the sky itself and clouds.

The combination of direct beam and diffuse radiation is called the global radiation; the level of global radiation varies through the day because the path length the light has to take through the atmosphere is greatest when the sun appears low in the sky in the early morning and late afternoon and least with the sun overhead at solar noon. For the same reason there are variations with both season and latitude. The total solar radiation received in a day can vary from about 0.5 kWh/m² on a sunny northern winter day to over 6.0 kWh/m² on a sunny day in the tropics. Obviously on a cloudy day the solar energy reaching ground can be almost zero. On a really clear, cloudless sunny day diffuse solar energy reaching the ground may be 15 to 20% of the global irradiance, but on an overcast day it can only be 100%.

Because of the variability of solar irradiation, the performance of solar pumping systems can be strongly influenced by location, time of year and weather. However, just like with wind energy, solar energy tends to be consistently available at any particular location from one year to the next; in other words, analysis of past solar energy records allows accurate prediction of solar energy and hence of system sizing. To some extent it is easier to achieve accuracy with solar energy prediction than with wind energy prediction since it is less affected by local site conditions, (providing the site is not actually shaded by trees), and any errors in solar energy measurement have less influence on system performance than comparable percentage errors in wind speed recording.

4.8.4 Performance Estimation

Accurate sizing of a solar pump array is necessary to minimize the array size for a given duty and thereby to achieve the most cost-effective system possible. If there is doubt about the sizing, a technically acceptable approach is to use a larger than necessary system in order to guarantee an adequate output, but the cost will tend to increase in proportion with power rating; 10% over-size therefore means 10% more expense than necessary.

Solar pumps used for irrigation should be sized for the "critical month"; that is when the system is most heavily loaded in relation to energy available. This is usually the month of maximum irrigation water demand, which fortunately is rarely one of the least sunny months, since crop irrigation water demand and solar energy tend to be well correlated. By contrast, a solar pump for drinking water supply duties is likely to need to deliver a constant daily water output, which would in this case make the least sunny month the critical month for sizing purposes.

Fig. 138 gives an example of mean monthly solar energy availability plotted on the same page as mean monthly gross irrigation water requirement for the Lake Tchad region of Africa. It can be seen from this that the month of June is not the sunniest month, but it is the critical month in terms of solar irrigation pump sizing, for that is when the water demand is largest in relation to the energy availability. The sunniest month, April, is not a critical month as it is too early in the growth cycle for the crop in question (i.e. cotton), while the fields are fallow in the second sunniest month of March.

Determination of the critical month, the mean daily water demand and the mean daily solar irradiation for that month are the starting points to size a solar pump (and hence to estimate its likely cost). Irradiation figures can be obtained from most national meteorological departments, and they are also published covering the whole world, for each month in references such as [57] [59] (as well as in some of the brochures of several of the larger manufacturers of solar photovoltaic power systems). Such published data are more region-specific than location-specific so there is a good chance of making an adequately accurate judgement simply on the basis of regional irradiation figures.

References such as [57] and [60] give a detailed explanation of rigorous methods for solar photovoltaic system sizing, but for most purposes a reasonable estimate can be made using a rule-of-thumb approach, such as the one explained below. In any case, most photovoltaic system manufacturers maintain computer models which they use for sizing systems; these generally include a world-wide data-base, so when requested to provide a quotation they can minimize the system size so as to price competitively. Therefore the potential purchaser should request quotations from several sources and compare the sizings and prices quoted. Care should be taken, however, to avoid simply taking the cheapest system and thereby possibly obtaining one which is undersized for the duty required.

Solar pump economics are discussed in more detail in Chapter 5, but in general terms, solar pumps begin to be economically competitive for applications where:

a. the peak daily head-flow product is under 150m³m (eg. 60m³/day through 2.5m head)

and where

b. the mean daily irradiation is greater than 4.2 kWh/m² (or 15MJ/m²) in the critical month.

A rule-of-thumb method may be used to obtain an approximate idea of the size and cost of a photovoltaic array, as follows:

i. estimate the peak daily hydraulic output (in kWh) required, using either Fig. 13 or the following formula:

$$E_{hyd} = \frac{Q\,H}{367} \text{ kWh/day}$$

where Q is the output in m³/day and H is the head in metres. eg. 8mm of water on 0.3ha pumped through a head of 10m; the volume is 24m³/day which requires (24 x 10)/367 = 0.654kWh(hyd)/day;

ii. assume a figure for the sub-system efficiency; i.e. the efficiency of conversion of electricity to hydraulic output (wire to water).

This is sometimes given in manufacturers' brochures or in technical studies such as [55] or [56]. However for small low head (2-5m) systems a figure of around 30% may be appropriate, while 40% can be achieved by more powerful and higher head units (say 5-20m head);

iii. divide the daily hydraulic output by the efficiency (say 35%) to arrive at the daily electrical energy demand for the system; eg. in the same example, 0.65kWh/0.40 = 1.625kWh(e) (assuming a sub-system efficiency of 40%);

iv. refer to the map in Fig. 139 and note the approximate Clearness Index for the location in question; eg. if we take Dakar in Senegal, West Africa, the map shows Dakar in the zone between 60 and 70%, at say 63%. The latitude of Dakar is approximately 15°N, so reference to the mean global extra-terrestrial radiation chart in Fig. 139 gives a mean of 9.73kWh/m². This needs to be multiplied by the relevant Clearness Index of 63%, which yields 0.63 x 9.73 = 6.1 kWh/m² per day on average. To allow for below average months and the inaccuracy of this estimating technique it is prudent to reduce this value by 20% for solar pump sizing purposes to give, say 0.8 x 6.1 = 4.9kWh/m². To summarise this; take the extraterrestrial irradiation for the latitude, multiply by the Clearness Index given on the map, and finally reduce by 20%.

v. Take the daily electrical energy requirement (calculated in (iii) above) and divide it by the daily irradiation (as calculated in (iv) above) and multiply the result by 1,200 to get the approximate peak watt rating required for the array. In the above example it is (1.625/4.9) x 1200 = 398W(p).

vi. Finally, arrays come in nominal 35 or 40W(p) modules, so it is then necessary to divide the answer by 35 or 40 (as appropriate) and round up the answer to an integer (whole) number of modules. e.g., with 40W modules we get 398/40 = 9.9, therefore 10 x 40W(p) modules would be needed, giving a true array rating of 400W(p).

Typical total total solar pumping system costs (1985) can be estimated on the basis of US $15-25 per W(p) of array, depending on the supplier and the specification of the system. Lower powered systems in the region of 200 to 500W(p) will tend towards the upper end of this price range, while higher powered systems, of say 2,000W(p) or more will tend towards the lower end. These figures are of course very approximate, as is the sizing methodology outlined, but with the example worked through we can see that the above example requiring a 400W(p) system, is likely to cost in the region of US $6,000-8,000. These costs are likely to come down during the next few years by perhaps 25 to 50%, as the volume of photovoltaic manufacturing increases.

4.9 HYDRO POWER

4.9.1 Background and State-of-the-Art

i. **The use of water power for irrigation**

The main shortcoming of water power as an energy resource for irrigation is that it is only available for convenient use in a limited number of locations having suitable flows and heads to engineer an effective site. Also, most regions with hydro-potential tend also to have plentiful rainfall, which often makes irrigation a low priority as a hydro-power application. However, despite these drawbacks, there are many areas where rainfall is seasonal (eg. where water can usefully be applied during the hot dry summer months to gain an extra crop or to increase yields and, similarly, in mountainous areas, where snow-melt water often provides some limited hydro-potential in the dry summer season). However, steep hydraulic gradients in hilly country generally allow any valley floor areas to be gravity irrigated by digging contoured canals, which take water from a stream and direct it along the side of a valley at a lesser hydraulic gradient then the main stream. This practice is of course widespread in much of the central Asian massif. When gravity does the work, pumping techniques are generally not needed.

Therefore, the main applications for the use of water power to pump water will be in lowland areas, inevitably using low head water drops as a power source, to irrigate land which would not be accessible to gravity water flow. An obvious important application, common in China, is to extend the command area available to a gravity irrigation scheme to take in higher land above a dam. In some arid regions, or regions with dry seasons, there are large perennial rivers where the river current can be utilized to lift water which would otherwise flow past parched fields. Even in some wetter and more mountainous areas there are situations where water power could allow irrigation of terraces or plateaux that are inaccessible to gravity flow; this can be of importance where flat land with soil suitable for cultivation is scarce.

Given a suitable site in proximity to a suitable need, hydro-power has a number of important and fundamental attractions:

a. it is generally available 24h/day

b. it is a relatively concentrated energy source

c. the available energy is easily predictable in areas where river flow data are available

It follows from these points that hydro-powered systems tend to have a high power to size ratio and hence a favourable power to cost ratio, they also tend to be mechanically simple and robust, and therefore have long working lives and require limited and simple maintenance. As a result hydro-power can be one of the most economic sources of power for those fortunate enough to have a suitable resource available.

j.i.   General principles

All water power applications involve removing energy from falling or flowing water. The power P available from a flow of Q litre/sec falling through a head of H m is:

$$P = 9.8 \times Q \times H \quad \text{watts}$$

Where 9.8 is the gravitational constant (in metric units of m/s^2), and if Q is in m^3/s, then the above formula gives P in kW. In otherwords there is 9.8kW available for a flow of $1m^3/s$ per meter drop. The actual output will be reduced by multiplying by the system efficiency; eg. a device with an efficiency of 50% will convert half the available power.

The efficiency of hydro-powered systems is high. A good small turbine will typically be 70% efficient and even water wheels and other more primitive devices tend to be 30 to 60% efficient, as indicated in Table 21.

Table 21   EFFICIENCY OF HYDRO POWERED SYSTEMS

type of device	efficiency factor (typical)
Undershot water wheel	0.30-0.40
Vertical shaft water mill	0.20-0.35
Poncelet undershot or Breast wheel	0.50-0.65
Overshot water wheel	0.50-0.70
Impulse turbine (eg Pelton)	0.70-0.85
Reaction turbine (eg Francis, Prop.)	0.60-0.80
Water pressure engine	0.60-0.80
Turbine-pump	0.35-0.50
Hydraulic ram (hydraulic output)	0.30-0.60
River current converter	0.25-0.30

Note that in all cases above, except turbine-pumps and hydraulic ram pumps, we are considering the conversion simply of water power to shaft power; to pump water then requires the addition of a pump or water lifting device which in turn will have further inherent inefficiencies. The exception has an efficiency figure relating to the hydraulic output divided by the hydraulic input, since there is no measurable intermediate production of shaft power.

It can be seen that the two key parameters required to estimate the power potential of a water resource are head and flow. Techniques for doing this are discussed in most references on micro-hydro power, such as [61] and [62]. The static head is the most straightforward to measure; methods which can be used include:

-   surveyor's levelling equipment;

-   spirit level, wooden pegs and straight edge;

-   a length of hosepipe with a pressure gauge (1m of water = $0.1kg/cm^2$ or .098bar or or 1.42p.s.i. or 9.81kPa);

the hosepipe is completely filled with water with its open end submerged in the upper water source and the other end fitted with the pressure gauge is held at the lower level to measure the actual static head;

- barometer/altimeter (for higher head applications); (i.e. surveying barometer/altimeter)

- accurate existing large-scale maps to take advantage of a previous survey

The flow rate can vary a lot from season to season; obviously the key period of interest to farmers will be the irrigation season. With larger rivers, especially those having existing hydro-electric schemes on them, discharge measurements will already have been carried out and been recorded and will usually be available from the relevant river authority. However, in most cases involving small-scale equipment for hydro-powered water lifting it will only be of interest to consider quite small streams, rivers or canals, for which no data will be available. There is a variety of methods of stream gauging varying in sophistication and accuracy, such as:

a. dam the stream and measure the overflow by measuring the time to fill a container;

b. construct a small rectangular weir from wood or concrete and use the method given in Fig. 140

c. in streams where it is not practicable to build a weir, measure the depth of the stream at equal intervals across its width (at a point where the stream is straight and uniform) as in Fig. 141. Then time a float drifting down the centre of the measured section. This gives the speed of the current at the centre of the stream; the mean velocity will be 0.6-0.85 of this figure. A rough, rocky stream bed requires a factor of 0.6, while smooth muddy surfaces require a factor of 0.85. The flow is calculated from the product of the mean velocity and area of stream cross-section using appropriate units:
    e.g. velocity in m/s. x cross-section in m²
    gives flow rate m³/s.

Obviously a current meter or a ship's log, if available, can more conveniently be used to measure the current velocity instead of timing a float.

d. dilution salt gauging is a new technique in which a known quantity of salt is poured into the centre of the stream at one point. An electrical conductivity meter is then used downstream from that point to measure the dilution. The data can then be used to determine the flow rate;

e. use of automatic stream gauging instruments; a calibrated weir is used to derive the day by day river flow on an electric or clockwork water level recorder.

In some cases it may be of interest to use river or canal current powered devices, in which case it is necessary to measure the current velocity

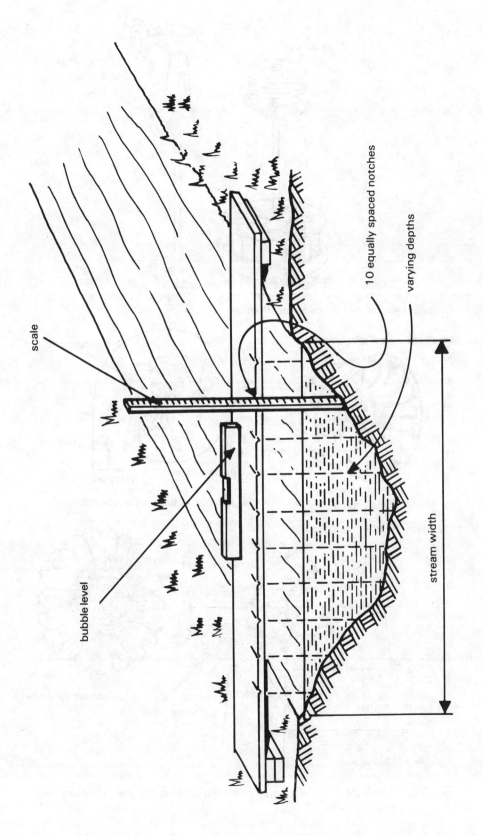

Fig. 141 Method of stream gauging without the need to build a weir

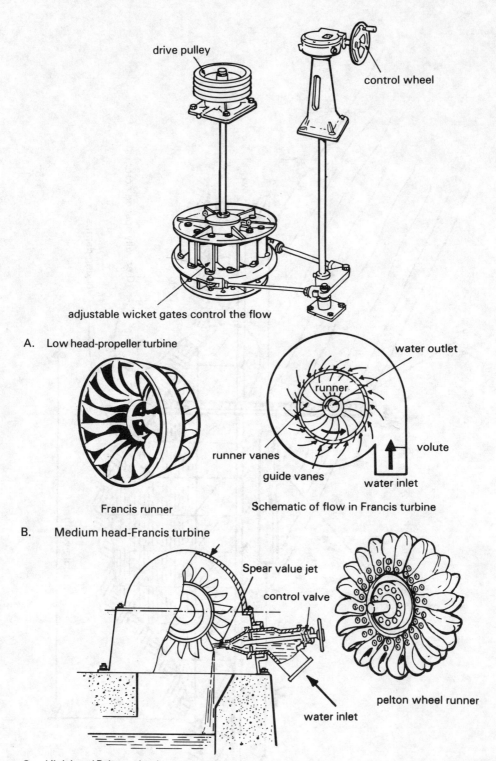

Fig. 142 Some of the main types of hydro turbine for low, medium and high heads

either by timing the passage of a float thrown into the current between two points on land which can be aligned with it as it floats past, or a calibrated current metering device (or ship's log) can be used.

It is extremely difficult to estimate flow or current velocity by eye; human judgement is peculiarly bad at assessing these kinds of factors and very inaccurate results are likely.

4.9.2 Use of Turbines for Water Lifting

i. General types of turbine

As with pumps, there are families of different types of turbine to deal with different types of situation. Briefly, these can be characterized as follows:

a. low head   -   propeller/Kaplan

b. medium head   -   Banki/Francis

c. high head   -   Pelton/Turgo

A typical small fixed blade propeller turbine is shown in Fig. 142 A. These are only adequately efficient over a narrow range of flows. Where variable flow and power is needed, adjustable gates are provided and the turbine runner may have fixed or adjustable pitch blades. The latter is known as a Kaplan turbine and is more efficient over a wider range of flows than a fixed pitch propeller turbine. But the complication of adjustable pitch runner blades is expensive, and therefore is only normally applied for larger-scale installations.

There is a middle head range of turbines, some of which involve mixed flow, rather like a mixed flow pump in reverse. Turbines of this kind run full of water, and are known generically as Reaction Turbines. The most widely used medium-head reaction turbine resembles a centrifugal pump (except the flow travels radially inwards from a spiral casing) and is a Francis turbine; Fig. 142 B. Reversed centrifugal pumps have actually been used as a cheap form of turbine, and can be quite acceptably efficient over a narrow range of flows and speeds at a given head. Another type of medium head turbine is known as the Banki (alias Ossberger, Mitchell or Cross-Flow) turbine. Here a jet of water impinges on a set of curved blades mounted between two discs, travels through the centre of the rotor and emerges from the far side again. Turbines of this kind do not run full of water and are known as impulse turbines in that they derive their rotation from deflecting a jet of water. High head applications are generally dealt with by another type of impulse turbine known as a Pelton Wheel (Fig. 142 C). With this a high speed jet of water is directed at a series of buckets set around the rim of the wheel. Each bucket has a central splitter which splits the jet in two and directs it almost back in the direction it came from and outwards from the rim of the wheel.

The reasons for using different turbine types at different heads and flows relate to the efficiency and speed of rotation that is required. Low head types of turbine, like propeller turbines have a high specific speed; this means they tend to rotate faster in relation to the velocity of the water travelling through them, which is important at low heads if electrical machinery, or rotodynamic pumps, are to be driven. However high heads would cause equipment of this kind to run at unacceptably high speeds, (and other problems would also occur) so lower specific speed types of turbine become necessary as the head increases. The interested reader should consult a standard text book on hydro-turbines as further discussion is beyond the scope of this work.

The shaft of a turbine can be connected directly to a generator, which then can be used to power electric pumps for irrigation, or it can be directly coupled to an appropriate centrifugal or other rotodynamic pump. The route via electricity is of interest in that irrigation water pumping is highly seasonal, and a system which produces electricity can in many cases perform useful duties other than irrigation, such as providing light at night, post harvest processes, etc. However electricity generation inevitably involves a higher level of engineering sophistication and investment than is inherent in powering a simple pump by a simple turbine. Also, any hydro-electric plant needs to be well protected from flooding, while a simple turbine powering a pump is much less liable to damage and can therefore be more simply installed. There are also losses of efficiency inherent in converting shaft power to electricity, transmitting the electricity and then converting it back to shaft power; this could absorbe 25 to 40% of the energy converted.

The chosen approach must depend on the size of the hydraulic resource, and the potential for satisfying other economic applications, plus of course the financial and organizational resources of the user community.

ii.   Chinese turbine-pumps

The Chinese have taken the combination of turbines and pumps to the logical conclusion of producing a large range of integrated turbine-pump units. Although this is a relatively recent development (originating in the early 1960s) some 60,000 turbine pumps were reportedly in use [63] irrigating 400,000ha by 1979.

Chinese turbine-pumps are generally for low head applications, where the hydro-power source will often be a canal jump in an irrigation scheme or a weir on a canalized river giving a head in the region of 1 to 15m. Therefore the turbine most commonly used is a fixed pitch propeller turbine which is appropriate for low heads. This is generally mounted with a vertical shaft surrounded by fixed gates; it therefore tends to be at its most efficient only over a narrow range of flow rates. A centrifugal pump impeller is mounted on the same shaft as the turbine, back-to-back as in Fig. 143. Where a high head water supply is needed, multi-stage centrifugal pumps may be connected to the turbine as in Fig. 144. In some cases an extension drive shaft can be fitted, as in Fig. 145; this allows the turbine to be used as a general power source at times when there is no demand for irrigation water. For example it can readily be applied to powering a small rice mill, oil expeller, generator, etc. This clearly can greatly enhance the economic value to be gained from the installation.

A large variety of different sizes and models of turbine pumps are made by numerous small manufacturers in China, and attractively low prices have been quoted for the export market. Table 22 indicates the range of sizes of single stage turbine pumps manufactured by a typical production unit, the

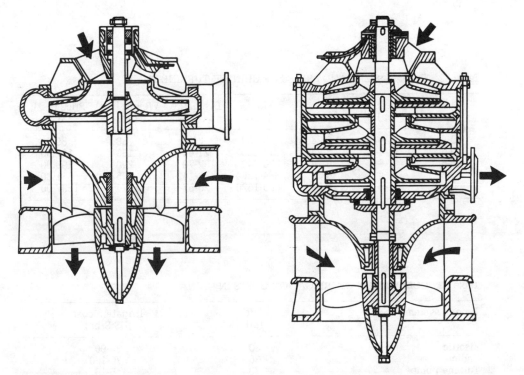

Fig. 143 Single stage high lift turbine pump (cross-section)

Fig. 144 Multi-stage high lift turbine pump (cross-section)

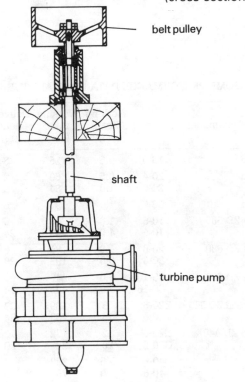

Fig. 145 Turbine pump fitted with extension drive shaft (show part-sectioned)

Table 22   TYPICAL SIZES AND PRICES OF CHINESE TURBINE-PUMPS

Type	Head (m)	Lift (m)	Flow (l/s)	Discharge (l/s)	Weight (net) (kg)	Price (net) (US $)
20-6	1-4	6-24	88-180	6-12	59	89.00
30-6	1-4	6-24	210-420	21-42	155	127.00
40-6	1-4	6-24	370-740	37-72	290	198.00
60-6	1-4	6-24	684-1354	70-142	985	571.00
60-16	1-6	16-96	650-1620	25-57	1374	1237.00

Table 23   COMPARISON OF IRRIGATION COSTS IN CHINA

Type of system	Lift (m)	Irrigation cost (US $/ha)
Electric	30	35-90
Engine	30	70-130
Turbine pumps	70	5-15

Table 24   SOME PERFORMANCE DATA OF SMALL TURBINE PUMPS

Turbine pump model		Working head (m)	Working flow (l/a)	Delivery head (m)	Delivery flow (l/a)	Shaft power (watts)	Overall efficiency (%)
High yield	10-0	0.5	15	3	0.8	47	32.8
High yield	20-4	0.5	60	2	7.8	203	
	20-6	1.0	84	6	6.0	570	42.8
High yield	30-6	1.0	190	6	16.0	1,340	50.6
Low head ZD 680	10-6	1.0	20	6	1.2	130	
	10-6	4.0	40	24	2.4	1,000	36.0
Low head ZD 680	20-6	1.0	81	6	6.4	560	
	20-6	6.0	198	36	15.5	8,180	47.0
Low head ZD 680	30-10	6.0	446	60	19.2	19,250	
	30-16	6.0	446	96	11.2	19,250	43.0
Medium head ZD 540	Z20-6	5	139	30	10	4,620	
	Z20-6	14	232	84	16	21,600	42.0
High head ZD 440	G20-6	12	172	72	11	14,400	
	G20-6	20	222	120	14	30,900	38.0
High lift	40-6	0.5	238	3	20.9		
	40-6	5.0	750	30	66.1		
High lift	40-17	1.0	333	17	3.3		
	40-17	6.0	815	102	8.5		

Youxi Turbine-Pump Plant, Fujian, and the (1981) export prices quoted for these by the Fujian Provincial Agricultural Machinery Import & Export Corporation.

Because it can be misleading to make judgements on the economics of Chinese equipment on the basis of Chinese export prices converted at the prevailing exchange rate, some limited Chinese data on the relative costs of alternative pumping systems within the Chinese economy are of interest. It can be seen from Table 23 that turbine pumps are significantly cheaper than either electrically energized pumps or engine pumps, as is to be expected. Whether the same differentials would apply elsewhere is open to question, but the turbine-pump is almost certain to be economically attractive wherever there is the need for irrigation water combined with suitable installation sites.

Table 24 indicates the performance of a selection of small low, medium and high head turbine-pumps, including estimates of shaft power and efficiency. The overall efficiency is in the range 32 to 50% for the models considered; this implies that the turbines and pumps considered individually rather than in combination have efficiencies in the 56 to 71% range, assuming roughly equal efficiency for each. Other models exist with claimed overall efficiencies as high as 58%.

Because fixed pitch propeller turbines only have a narrow operating range where high efficiency can be achieved, it is important that they are accurately sized to suit the flow and head. Where varying flow conditions occur, it is usual to install several small units rather than one large one. This means each unit can always be run near to its optimum flow condition by shutting them down one by one to cater for reduced flows. A good design strategy is to install two units, one with twice the flow capacity of the other; then both are used under the maximum flow condition, the larger can take $\frac{2}{3}$ of the maximum flow on its own, and the smaller can take $\frac{1}{3}$ of maximum flow on its own. This allows efficient operation at $\frac{1}{3}$, $\frac{2}{3}$ and full-flow.

Turbine pumps are typically installed on a concrete platform built into a weir, as in Fig. 146. Therefore, although the turbine-pump unit is inexpensive, depending on the site, civil workings are likely to represent the largest cost-element. Pipes also will be expensive, but then for a given flow they will be equally expensive regardless of the choice of pumping system.

Higher heads or higher flows are commonly catered for by connecting turbine pumps in series or parallel.

4.9.3 The Hydraulic Ram Pump (or Hydram)

The hydraulic ram pump, or hydram, concept was first developed by the Montgolfier brothers, better remembered for their pioneering work with hot-air balloons, in France in 1796. Essentially, a hydram (shown schematically in Fig. 147), is an automatic pumping device which utilizes a small fall of water to lift a fraction of the supply flow to a much greater height. In other words, as with the turbine pump, it uses a larger flow of water falling through a small head to lift a small flow of water through a higher head. The main virtue of the hydram is that it has no substantial moving parts, and is therefore mechanically extremely simple, which results in very high reliability, minimal maintenance requirements and a long operational life.

Its mode of operation depends on the use of the phenomenon called water hammer and the overall efficiency can be quite good under favourable circumstances. Fig. 147 illustrates the principle; initially the waste valve (1) will be open under gravity, and water will therefore flow down the drivepipe (2) from the water source (3) (having been drawn through a strainer (4) to prevent debris entering the hydram). As the flow accelerates, the hydraulic pressure under the waste valve and the static pressure in the body of the hydram (5) will increase until the resulting forces overcome the weight of the waste valve and it starts to close. As soon as it starts to close, and the aperture decreases, the water pressure in the valve body builds up rapidly and slams the waste valve shut. The moving column of water in the drivepipe is no longer able to exit via the waste valve so its velocity must suddenly decrease; this continues to cause a considerable rise of pressure which forces open the delivery valve (6) to the air-chamber (7). Once the pressure in the air chamber exceeds the static delivery head, water discharges through the delivery pipe (8). Air trapped in the air chamber is simultaneously compressed to a pressure exceeding the delivery pressure. Eventually the column of water in the drivepipe comes to a halt and the static pressure in the casing then falls to near the static pressure due to the supply head. The delivery valve will then close, due to the pressure in the air chamber exceeding the pressure in the casing. Water will continue to be discharged through the check valve (9), after the delivery valve has closed, until the compressed air in the air chamber has expanded to a pressure equal to the delivery head. At the same time, as soon as the delivery valve closes, the reduced pressure in the casing of the hydram allows the waste valve to drop open, thereby allowing the cycle to start again.

The air chamber is a vital component, as apart from improving the efficiency of the process by allowing delivery to continue after the delivery valve has closed, it is also essential to cushion the shocks that would otherwise occur due to the incompressible nature of water. If the air chamber fills with water completely, not only does performance suffer, but the hydram body, the drivepipe or the air chamber itself can be fractured by the resulting water hammer. Since water can dissolve air, especially under pressure, there is a tendency for the air in the chamber to be depleted by being carried away with the delivery flow. Different hydram designs overcome this problem in different ways. The simplest solution requires the user to occasionally stop the hydram and drain the air chamber by opening two taps, one to admit air and the other to release water. Another method on more sophisticated hydrams is to include a so-called snifting valve which automatically allows air to be drawn into the base of the air chamber when the water pressure momentarily drops below atmospheric pressure at the moment after flow commences when the waste valve reopens. It is important with such units to make an occasional check to see that the snifting valve has not become clogged with dirt and is working properly.

This cycling of the hydram is timed by the characteristic of the waste valve. Normally it can be either weighted or pre-tensioned by an adjustable spring, and an adjustable screwed stop is generally provided which will allow the maximum opening to be varied. The efficiency, which dictates how much water will be delivered from a given drive flow, is critically influenced by the valve setting. This is because if the waste valve stays open too long, a smaller proportion of the throughput water is pumped, so the efficiency is reduced, but if it closes too readily, then the pressure will not build up for long enough in the hydram body, so again less water will be delivered. There is often an adjustable bolt which limits the opening of the valve to a

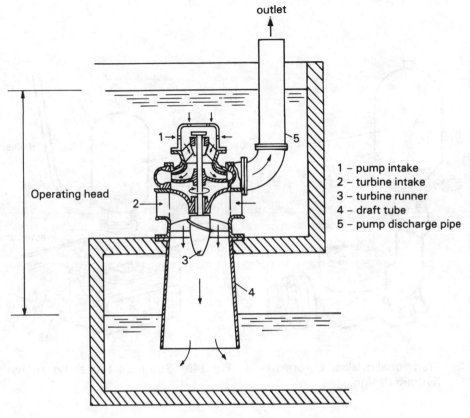

Fig. 146 Typical turbine pump installation

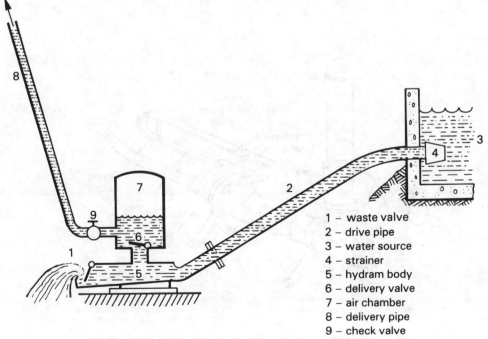

1 – waste valve
2 – drive pipe
3 – water source
4 – strainer
5 – hydram body
6 – delivery valve
7 – air chamber
8 – delivery pipe
9 – check valve

Fig. 147 Schematic diagram of hydram installation

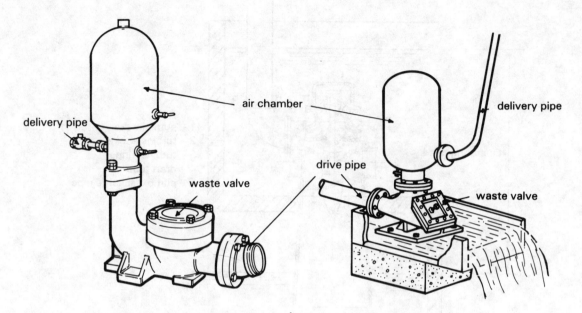

Fig. 148 Traditional (Blakes) European hydram design

Fig. 149 South-East Asian type of hydram

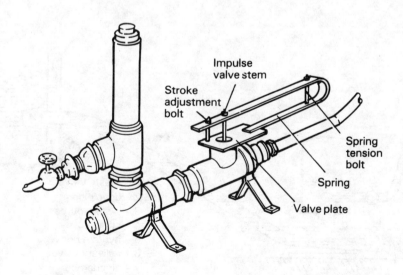

Fig. 150 Low-cost hydraulic ram using standard pipe fittings

predetermined amount which allows the device to be tuned to optimize its performance. A skilled installer should be able to adjust the waste valve on site to obtain optimum performance for that particular hydram and site.

Therefore, it can be seen that the output of a hydram is constant, 24 hrs/day, and cannot readily be varied. A storage tank is usually included at the top of the delivery pipe to allow water to be drawn in variable amounts as needed.

i. **Installation requirements**

In a typical hydram installation, a supply head is created either by digging a small contoured diversion canal bypassing a river, or in some cases, particularly with small streams, it is normal simply to create a weir and to install the hydram directly below it.

Where greater capacity is needed, it is common practice to install several hydrams in parallel. This allows a choice of how many to operate at any one time so it can cater for variable supply flows or variable demand.

The size and length of the drivepipe must be in proportion to the working head from which the ram operates. Also, the drivepipe carries severe internal shock loads due to water hammer, and therefore normally should be constructed from good quality steel water pipe. Normally the length of the drivepipe should be around three to seven times the supply head. Also, ideally the drivepipe should have a length of at least 100 times its own diameter (eg. the design length for a 100mm (4in) drivepipe is about 10m, and for a 150mm (6in) drivepipe it is about 15m). The drivepipe must generally be straight; any bends will not only cause losses of efficiency, but will result in strong fluctuating sideways forces on the pipe which can cause it to break loose.

The hydram body needs to be firmly bolted to a concrete foundation, as the beats of its action apply a significant shock load. It should also be located so that the waste valve is always above flood water level, as the device will cease to function if the waste valve becomes submerged. The delivery pipe can be made from any material capable of carrying the pressure of water leading to the delivery tank. In all except very high head applications, plastic pipe can be considered; with high heads, the lower end of the delivery line might be better as steel pipe. The diameter of the delivery line needs to allow for avoiding excessive pipe friction in relation to the flow rates envisaged and the distance the water is to be conveyed. It is recommended that a hand-valve or check-valve (non-return valve) should be fitted in the delivery line near the outlet from the hydram, so that the delivery line does not have to be drained if the hydram is stopped for adjustment or any other reason. This will also minimize any back-flow past the delivery valve in the air-chamber and improve the efficiency.

ii. **Choice of hydram design**

Traditional hydram designs, such as Fig. 148, developed a century ago in Europe, are extremely robust. They tend to be made from heavy castings and have been known to function reliably for 50 years or more. A number of such designs are still manufactured in Europe and the USA in small numbers. The hydram in Fig. 148 differs from the schematic diagram of Fig. 147 in having its waste valve on the same side as the drive pipe, but its principle of operation is identical.

Lighter designs, fabricated using a welded sheet steel construction (Fig. 149) were developed first in Japan and are now in production in other parts of SE Asia including Taiwan and Thailand. These are cheaper, but only likely to last a decade or so as they are made from thinner material which will probably eventually corrode; nevertheless they offer good value for money and are likely to perform reliably for a respectably long time. However, hydrams are mostly intended for water supply duties, operating at higher heads and lower flow rates than are normal for irrigation. Therefore it is likely that the most useful hydrams for irrigation purposes will be the larger sizes having 100-150mm (4"-6") drivepipes. Some simple designs that can be improvized from pipe fittings have also been developed by aid agencies, (such as in Fig. 150) and some interesting versions have also been quite crudely improvized using scrap materials, such as in southern Laos using materials salvaged from bombed bridges and with old propane cylinders as air chambers. Needless to say, such devices are very low in cost; the pipes in the end cost considerably more than the hydram. They are not always as reliable as traditional designs, but nevertheless are usually quite adequately reliable and are easy to repair when they fail.

iii. Performance characteristics

Table 25 (from [64]) indicates the input capacity of different sizes of hydram; it is this input which determines the delivery flow at a given delivery head. The lower limit indicates the minimum input flow required for practical operation, while the upper limit represents the maximum possible flow a hydram can handle. Table 26 [64] then indicates the probable litres pumped in 24 hrs for each litre/min of drive water.

Although the costs of hydrams are apparently low, as soon as high flow rates are needed at lower heads, the size of hydram and more particularly of drive pipe, begins to result in significantly higher costs. Therefore hydrams are best suited to relatively low flow rates and high head applications, (perhaps terraced tree nurseries in mountainous regions) while a turbine-pump, as discussed in the previous section, appears more attractive for the lower heads and high flow rates that are more common for irrigation of commercial crops on low-land farms.

4.9.4 Water Wheels and Norias

The undershot waterwheel is probably both the most obvious and the oldest method of extracting energy from rivers. In many cases the device simply dips into the river and is turned by the movement of the current; see Fig. 151. In the example illustrated, from Vietnam, the entire structure is made of bamboo and bamboo tubes with one end closed are mounted around the rim of the wheel. The bamboo tubes dip into the river and re-emerge filled with water, which they to near the top, where the water pours out into a trough. Devices of this kind are quite widely used in SE Asia, including China, Japan and Thailand as well as Vietnam, and are known as "norias". Fig. 152 shows a Chinese version and illustrates the principle and method of construction quite clearly. The noria is similar in many ways to the Persian wheel and was discussed earlier in Section 3.4.1.

Table 25 HYDRAM INPUT CAPACITY

Nominal diameter of drive pipe	ins. bore		$1\frac{1}{4}$	$1\frac{1}{2}$	2	$2\frac{1}{2}$	3	4	5	6	7	8
	mm. bore		32	40	50	65	80	100	125	150	175	200
Volume of driving water required to operate the ram	litres per minute	from	7	12	27	45	68	136	180	364	545	770
		to	16	25	55	96	137	270	410	750	1136	1545

Table 26 HYDRAM PERFORMANCE

Supply Head (m)	Delivery head (m)											
	5	7.5	10	15	20	30	40	50	60	80	100	125
1.0	144	77	65	33	59	20	12					
1.5		135	96	70	54	36	19	15				
2.0		220	156	105	79	53	33	25	20	12		
2.5		280	200	125	100	66	40	32	24	16	12	
3.0			260	180	130	87	65	51	40	27	18	12
3.5				215	150	100	75	60	46	32	20	14
4.0				255	173	115	86	69	53	36	23	16
5.0				310	236	155	118	94	72	50	36	23
6.0					282	185	140	112	94	64	48	34
7.0						216	163	130	109	82	60	48
8.0							187	149	125	94	69	55
9.0							212	168	140	105	84	62
10.0	Litres pumped in 24 hours per l/min of drive water.						245	187	156	117	93	69
12.0							295	225	187	140	113	83
14.0								265	218	167	132	97
16.0									250	187	150	110
18.0									280	210	169	124
20.0										237	188	140

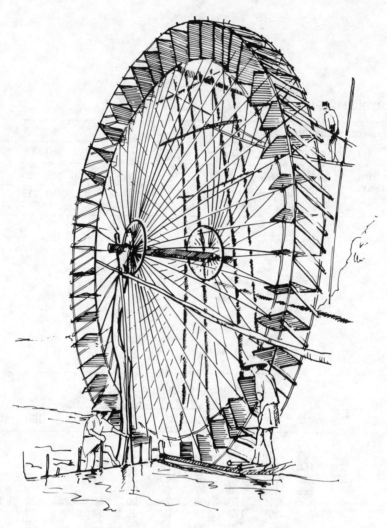

Fig. 151   Bamboo water wheels, Vietnam

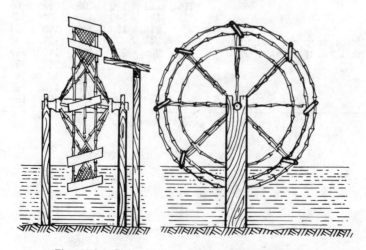

Fig. 152   Chinese type of small-scale Noria

The biggest shortcoming of the noria is that they need to be of a diameter somewhat greater than the head; this makes them fine for low head applications, but they get large and cumbersome for higher lifts. The example illustrated in Fig. 151 is 10m in diameter and is claimed to be able to irrigate about 8ha [65]. There is a small weir just visible in the illustration which creates a head of about 100mm at the base of the wheel which significantly improves the performance. The 10m diameter Vietnamese noria turns at the rate of about 1 revolution in 40 seconds and delivers water, typically, at the rate of 7 litre/sec. The cost of these Vietnamese norias is quoted as being the equivalent of US $225-450, [65]. Although the noria is attractive in being relatively inexpensive and also being capable of manufacture in the village, the sites where they may be used are limited and they are particularly prone to damage by floods. Therefore annual repair costs quoted for Vietnamese norias [65] can be as much as 30-50% of the capital cost of the installation. However, under Vietnamese conditions the same source indicated that some sample installations which were surveyed produced a return for the users in terms of value of grain production in the range 24-60% over the estimated costs.

A modern version of the Asian noria is the floating coil pump, versions of which have been tested by the Sydfynsgruppen (supported by DANIDA) at Wema on the Tana River in Kenya, by the Danish Boy Scouts on the White Nile at Juba in southern Sudan and by the German agency BORDA on the River Niger near Bamako in Mali [66] (see Fig. 153). The principle of the coil pump is explained in Section 3.6.4 and also illustrated in Fig. 50. These recent experimental river current powered irrigation pumping systems consist of floating undershot water wheels (mounted on floating pontoons made from empty oil barrels), and by using a coil pump it is possible to use quite a small diameter water wheel and to use it to lift water to a height considerably greater than its diameter (around 10-15m is possible from a 2-3m diameter rotor).

The claimed performance of the prototype floating coil pumps was 6.6 litre/sec against a delivery head of 5m with a river current velocity of 1.2 m/s on the Tana River at Wema. For reasons which are not known, the similar unit tested on the Nile only pumped 0.7 litre/sec against 5m head with the same river current velocity of 1.2m/s. Quite high current velocities in the range from 1-2m/s (2-4 knots) are necessary for devices of this kind. The Kenyan machine was made mainly from glass-reinforced plastic (fibreglass) and cost US $4,000 in 1979, but the much less efficient Sudan prototype used steel and wood, with flexible plastic pipe for the coil pump and only cost $350 In both cases the considerable length of flexible pipe required accounted for about one third of the total cost.

The Royal Irrigation Department of Thailand has developed a similar floating undershot waterwheel device to those just described, but in this case the wheel is mechanically linked to a conventional piston pump by means of sprockets and chains. It requires a minimum current velocity of 1m/s with a river flow of at least $0.6m^3/s$ and it is claimed to be capable of pumping from 0.3-1.5 litre/sec to heads of 60m-15m respectively. The cost was quoted at $1,450, not including the delivery piping and header tank.

A general problem with water wheels is that if the drive is taken from the main shaft, the costs become high in relation to the power available, because large slow-moving (and therefore expensive) mechanical drive components are needed to transmit the high torque involved. Also, with shafts only turning at between 1-5 rpm (which is typical of waterwheels) either very large pump swept volumes are required or a lot of gearing up is necessary to drive a smaller pump at an adequate speed; either way the engineering is expensive in relation to the power. Therefore waterwheels, although apparently offering simple solutions, are not always as easy to adapt for powering mechanical devices such as pumps as might at first be expected.

4.9.5 Novel Water-Powered Devices

i. The Plata Pump

An unconventional alternative to a small water wheel for powering a small pump is the Plata Pump (Fig. 154). This device, invented in New Zealand, and some efforts have been made to commercialize it internationally, although it is not believed to be very widely used. In it, a series of small turbine rotors are mounted on a single shaft along the axis of a cylindrical duct, approximately 2.5m long by 0.5m in diameter. It is a bit like a multi-rotored propeller turbine, although there are no diffusers or other static blades to control rotation of the fluid. The shaft drives two opposed single-acting piston pumps via a crank. The Plata Pump is intended to be mounted in a low dam or weir, so that it slopes at a slight angle and water runs downhill through it. It is not supposed to run full of water, but works best when running around ⅔ to ¾ full; perhaps because when running full the flow rotates and thereby causes loss of efficiency.

The Plata Pump is designed to operate on heads of 0.25-1m; usually this is engineered by placing the Plata Pump at an appropriate angle on a stream bed and building up a weir with rocks or other material to create the necessary head. The overall efficiency of the Plata Pump has been measured as being in the 6-30% range at delivery heads from 6-90m; [66]. The best efficiency was recorded at 24m head. Typical performance with a working flow of 85 litre/sec is 1.3 litre/sec at 6m, 0.25 litre/sec at 24m, 0.11 litre/sec at 38m. A working flow of 153 litre/sec was necessary to allow 90m delivery head to be reached.

The price of the Plata Pump in kit form was about US $2,000 in 1980, [66], so it seems to be relatively expensive in relation to its performance in comparison, for example, with Chinese turbine pumps described earlier. It also appears to be less efficient and robust than the turbine pump.

ii. The river current turbine

The energy of motion (or kinetic energy) available in river or canal currents is, exactly as for wind, proportional to the cube of the velocity. The relationship between power and velocity is:

$$P = \tfrac{1}{2}\rho A V^3 \text{ watts}$$

where ρ is the density of water (1000kg/m³ for fresh water), A is the area of cross section of current in m² and V is the mean velocity through the cross-section in m/s. From this the following power densities can be calculated, as shown in Table 27:

Table 27   POWER DENSITY IN WATER CURRENTS
                  AS A FUNCTION OF WATER VELOCITY

velocity (m/s)	0.5	1.0	1.5	2.0
(knots)	1.0	2.0	3.0	4.0
power density (kW/m²)	0.06	0.5	1.7	4.0

It is interesting to compare this with Table 16 giving the equivalent result for wind; similar power is experienced in water at about 1/9 the velocity in wind needed to achieve the same power, due to the much higher density of water. Since a mean wind speed of 9m/s would be considered most attractive for the economic use of wind power, it is clear that even currents of 1m/s may be more than adequate as a power source if they could be exploited efficiently. Moreover, a major problem with exploiting wind energy is its great variability, but river currents generally flow steadily 24 hours per day, [67].

The traditional and most obvious technique, undershot water wheels, are an inefficient means to exploit currents, since the bulk of the machine is external to the water at any one time and therefore provides no power. Therefore the Intermediate Technology Development Group (UK) undertook a programme, financed by the Netherlands Government, to develop turbines which would operate efficiently entirely submerged to extract shaft power from river currents; (the author must declare an interest, since the device in question was developed under his technical supervision).

The main device tested is a vertical axis cross-flow turbine similar to the Darrieus windmill in principle (see Section 4.7.2). It has the advantage that the vertical drive shaft conveniently comes through the surface, so that any mechanical components can be located on the deck of a pontoon above water. Unlike a water wheel this device can intercept a comparatively large cross-section of current using little material other than a rotor to do so; moreover it turns relatively fast in relation to the current (13.5rpm in this case with a 1m/s current) which reduces the gearing needed to drive a pump at a reasonable speed. The concept is illustrated in Fig. 155, while an actual working irrigation pumping unit, installed on the White Nile near Juba in southern Sudan, worked reasonably reliably, was fitted with a 3m diameter vertical axis rotor (3.75m² cross-sectional area) powering a centrifugal pump via a two stage toothed belt speed increaser. It was tested and found to pump approximately 3.5 litre/sec through a head of 5m with a current of 1.2m/s. The rotor efficiency is 25 to 30% (as with a small windmill) and an overall system efficiency of 6% has been achieved, including pipe losses as well as pump and transmission losses. The prototype, which cost the equivalent of US $5,000 to build, has been used successfully to irrigate a 6ha vegetable garden. Smaller, simpler, low-cost versions were also tested near Juba. Costs are difficult to determine with prototypes, but the device appears to be potentially economically attractive.

The potential for using the river to pump its own water has been demonstrated, but further work will be needed to optimize this device and adapt it for commercial production. However, the considerable power potential in many river, canal and for that matter tidal currents should ensure that this at present little known and little thought of power source gains increasing recognition, and will eventually be made use of. This is particularly because many large rivers (such as the Nile, Euphrates, Zambesi, Indus), flow through regions which are arid or which have several months of dry season.

4.10   BIOMASS AND COAL (THE NON-PETROLEUM FUELS)

4.10.1 General Description

i.   The availability and distribution of fuels

The developed world runs mainly on petroleum oils because, even at today's prices, oil is a cheap fuel and is far more convenient to use, transport and store than any alternative. But, the world's supply of easily recovered petroleum is fast diminishing and we are running towards an era when supply will no longer exceed the demand. Already, many developing countries can no longer afford to import sufficient oil for their present needs, let alone to satisfy expanded future energy demand.

The main fuel alternatives to petroleum are solid fossil fuels such as coal, lignite and peat plus the "biomass fuels", which are derived from recently living rather than fossilized organic material. Estimates on world coal reserves vary, but there is general agreement that the amount of coal available is in the order of ten times the total for oil in energy terms, so although oil shortages due to physical depletion can be expected within decades, coal should remain available for centuries. In fact it is likely that a need to control atmospheric pollution will constrain the use of coal long before the reserves are exhausted.

The main problem with biomass as a fuel is its uneven distribution, being most available in forested regions where it is little needed. Considered as a global resource there is no shortage of biomass [68], since:

a.   the world's annual rate of use of energy is only about 10% of the annual natural rate of photosynthetic energy storage (i.e. solar energy stored in plant matter);

b.   stored biomass on the earth's surface at present is approximately equivalent to the entire proven fossil fuel reserves (oil, coal and natural gas).

Biomass, in the form of firewood, charcoal, agricultural residues, or dried animal dung is already the main energy resource for the majority of the poorer half of humanity, over 2,000 million people. They currently use 15% or more of total world energy, entirely in the form of biomass. Certain poorer countries depend on biomass fuels at present for over 90% of their energy needs (i.e. mainly for cooking fuel). It has been estimated [69] that 58% of energy use in Africa, 17% in Asia and 8% in Latin America is currently met from biomass resources. Therefore biomass is already a huge and vital economic resource, although it is usually informally exploited mainly for cooking purposes and still remains rarely used for the production of shaft power via heat engines, and even less so for irrigation pumping.

The main problem with all fuels, but particularly with biomass, is distribution. Even energy-intense and efficiently marketed commercial fuels like petroleum are often in short supply in many remote areas of developing countries, so it is not surprising that more bulky and less valuable biomass fuels present even greater problems of distribution. Fuelwood supplies in the Third World do not coincide at all well with population distribution, so that there are areas of scarcity and areas of surplus. In most cases the fuels in question are too bulky and low in commercial value to allow such a simple solution as transporting them from areas of surplus to areas of scarcity.

The size of the areas suffering from a deficit of biomass fuels is increasing due to population pressure, combined in some cases, such as the African Sahel region, with apparent climatic changes. There is a narrow gap between a sustainable rate of firewood foraging and the catastrophic rate of removal of vegetation which leads to desertification. Lack of firewood leads to increasing use of dung as a fuel (in India it is perhaps the primary fuel for cooking), and the use of dung for fuel rather than fertilizer leads to a further decline in soil quality and contributes to the same problem. It has been estimated [70] that almost 3 billion people will face cooking fuel shortages by the turn of the century. Despite this grave situation, most of the R&D on biomass utilization has focussed on its use as a petroleum substitute (eg. alcohol for powering cars in Brazil), to address the "oil crisis" of the rich, rather than the "wood crisis" of the poor. Some of these latter developments, involving the large scale production of biomass fuels, give rise for concern as to the impact they might have on the capability of countries to feed their populations. For example, Brown [71] indicates the land requirements in Brazil to fuel a car compared with those to feed a person, are as indicated in Table 28.

Table 28   LAND REQUIREMENTS IN BRAZIL TO PRODUCE GRAIN
FOR FOOD OR FOR FUEL ALCOHOL

grain crop use	quanitity required	cultivated land area
subsistence diet	180kg/yr	0.1 ha
affluent diet	700kg/yr	0.4 ha
medium sized car	2,800kg/yr	1.4 ha
large (US) car	6,600kg/yr	3.2 ha

This shows that even a medium sized car doing only 12,000km per year needs a land area to produce its alcohol fuel sufficient to feed 14 people on a subsistence diet. This issue of food versus fuel is an important one, but where the fuel crop is used for irrigation pumping to produce food crops, especially in a small-scale "on-farm" process, the same objections need not apply.

ii.   The nature and calorific value of fuels

All fuels involve the combustion of carbon and usually hydrogen with atmospheric oxygen to produce mainly carbon dioxide and water, plus heat. As far as heat production is concerned, the relative merits of various fuels are best summarised in terms of the calorific value of total heat released when they are burnt; see Tables 29 and 30, [6]. This can be expressed in energy per unit weight or per unit volume, but in the end, what matters most is the energy per unit cost. Also of importance with biomass fuels if grown as a fuel crop is their productivity, which obviously effects their cost-effectiveness. This must clearly depend on many factors such as location, soil quality, etc. The most productive areas for photosynthetically produced material are the forested regions; agricultural land is typically only half as productive per hectare in terms of biomass generation. Table 31, after Earl [72], shows some typical production rates, while Table 32 gives calorific values for a variety of fuels, both fossil and biomass. Fossil fuels tend to be more consistent in their properties then biomass. The calorific value of biomass fuels is particularly influenced by the moisture content of the fuel. For example, air dried wood, which normally has a moisture content of 25% will produce 50%-100% more heat than "green" freshly cut wood with a moisture content of 50%. Oven drying of wood increases the heat yield further (but requires expenditure of heat); fuel, once oven-dried, has to be kept in warm dry conditions before being burnt or it will reabsorb atmospheric moisture. Therefore there is a good case to be made for utilization of waste heat from any biomass engine system to help dry its own fuel supply.

The production rate of cultivated biomass materials is even more variable than its quality. Table 33 indicates examples of measured yields of potential fuel crops.

iii. **The potential for using biomass for fuelling irrigation pumps**

Biomass ought to be an ideal energy resource for irrigation, since the whole point of irrigation is to produce more biomass, usually for food rather than for fuel. A simple calculation confirms that it is possible, at least in theory, to grow more than enough biomass to fuel an engine and produce an additional food crop, even without considering using food-crop residues. For example, considering the irrigation system sized in Fig. 13, where 3ha is irrigated with 8mm of water on average per day and a pumping head of 10m; assuming that irrigation is necessary on 200 days per year, Fig. 13 indicates an average shaft power requirement of 13kWh/day or in this case 2,600kWh per year. If the power system can produce shaft power from fuel at 10% efficiency (which is possible at this power size) then the gross fuel requirement is for 26,000kWh or 94 GJ. It can be seen from Table 30 that this requirement could often be met simply from cereal crop residues for 3ha, but if a fuel crop was grown because, for example the residues were needed for other purposes such as cooking fuel, then just 0.1 to 0.2ha of eucalyptus (for example) would produce this fuel requirement. Therefore, the entire irrigation energy demand could be met from a fuel crop occupying in this example less than one tenth of the area to be cultivated. And 10m lift with 200 days per year irrigation is a more demanding irrigation energy need than would apply in many cases. Therefore, the reason biomass is not more often used for irrigation pumping may relate more to ignorance of this option or perhaps to lack of opportunity than to any technical constraint, or perhaps it is because the more knowledgable farmer who may know of this option is so often the more prosperous one who can in any case at present afford the greater convenience of using diesel power. The rest of this chapter sets out to review the biomass fuelled options and their advantages and disadvantages.

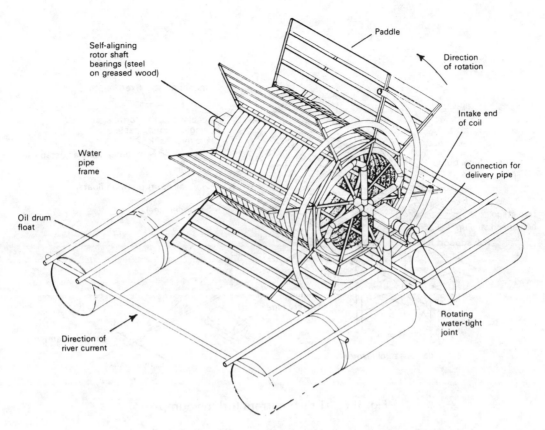

Fig. 153 Water wheel driven coil pump (see also Fig. 50)

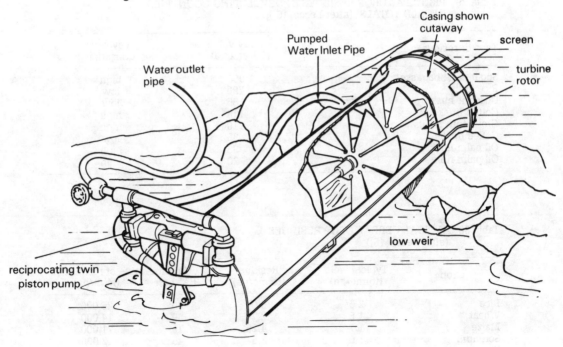

Fig. 154 Cut-away view showing general arrangement of a Plata pump installation

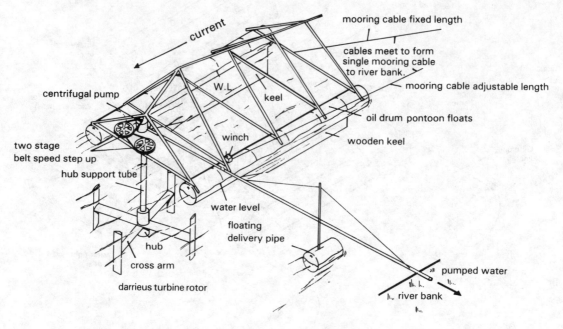

Fig. 155   IT river current turbine pump

Table 29   PRINCIPAL CROP RESIDUES IN DEVELOPING COUNTRIES (WORLD TOTALS) (after Leech, [6])

Type of residue	Energy value (GJ/yr x 10)	Level of utilization
Sugar cane bagasse	1,060	high
Rice hulls	790	low
Coconut husk	185	low
Cotton husk	110	high
Groundnut shells	110	high
Coffee husks	35	low
Oil palm husk	35	high
Oil palm fibre	20	high

Table 30   TYPICAL CEREAL CROP RESIDUES (after Leech, [6])

Crop	Typical yield (tonne/ha)	Residue (tonne/ha)	Calorific value of residue (GJ/ha)	(kWh/ha)
Rice	2.5	5.0	90	25,000
Wheat	1.5	2.7	49	14,000
Maize	1.7	4.25	76	21,000
Sorghum	1.0	2.5	45	12,000
Barley	2.0	3.6	65	18,000
Millet	0.6	2.0	36	10,000

Table 31 PHOTOSYNTHETIC CARBON PRODUCTION RATES
(after Earl, [72])

Type of land	Net primary production (tonne/ha)	Total world annual natural production (giga-tonne)
A. Forest		
Temperate deciduous	10	8
Conifer and mixed	6	9
Temperate rain forest	12	1
Tropical rain forest	15	15
Dry woodlands	2	3
Sub-total:		36
B. Non-Forest		
Agricultural	4	6
Grasslands	3	8
Tundra	1	1
Deserts	1	3
Sub-total:		18
Total:		54 Gt

Table 32 RELATIVE HEAT VALUE OF VARIOUS FUELS (APPROXIMATE VALUES)

	Calorific value/unit wt.		Calorific value/unit vol.	
	(MJ/kg)	(BTU/lb)	(MJ/m^3)	(BTU/ft^3)
A: Fossil fuels				
Petrol/gasoline	44	19,000	32,000	860,000
Fuel oil	44	19,000	39,000	1,050,000
Paraffin/kerosine	45	19,500	36,000	970,000
Diesel/gas oil	46	20,000	38,000	1,020,000
Coal tar/asphalt	40	17,000	40,800	1,100,000
Anthracite coal	35	15,000	56,000	1,500,000
Bituminous coal	33	14,000	42,900	1,150,000
Lignite (brown) coal	30	13,000	37,500	1,010,000
Peat	20	9,000	18,200	490,000
Coke	28	12,000	22,400	600,000
Natural gas (methane)	56	24,000	40*	1,020
Coal gas	9	4,000	20*	490
Propane (cylinder gas)	48	21,000	90*	2,400
Butane (cylinder gas)	47	20,000	120*	3,100
B: Bio-mass fuels				
Wood (oak)	18	8,000	14,400	390,000
Wood (pine)	20	9,000	10,000	270,000
Wood (acacia)	16	7,000	11,000	300,000
Charcoal	28	13,000	11,000	300,000
Sunflower stalks	20	9,000	10,000	270,000
Wheat straw	18	8,000	—	—
Beef cattle manure	14	6,000	—	—
Methanol (methyl alcohol)	20	8,600	19,000	500,000
Ethanol (ethyl alcohol)	28	12,000	28,000	700,000
Bio-gas (65% methane)	20	8,600	23*	600
Wood gas (typical) (producer gas)	—	—	5*	140
Vegetable oil	39	16,500	32,000	860,000

*Since these fuels are normally gaseous, the calorific value per unit is relatively low compared with liquid and solid fuels.

Table 33 POTENTIAL BIO-MASS VALUES OF SELECTED CROPS

Species	Location	Annual dry matter yield		Heat value		Tonne oil equiv. per ha./yr
		(ton/acre)	(tonne/ha)	(10⁶ BTU/acre/yr)	(GJ/ha/yr)	
Sunflower	USSR	13.5	30	200	530	12
Forage sorghum	Puerto Rico	30.6	69	460	1,210	28
Hybrid corn	USA (Miss)	6	13	90	250	6
Water hyacinth	USA (Fla)	16	36	240	630	14
Sugar cane (average)	USA (Fla)	17	39	260	680	16
Sugar cane (experiment)	USA (Cal)	32	72	480	1,250	29
Sudangrass	USA (Cal)	16	36	240	630	15
Bamboo	SE Asia	5	11	70	210	5
Eucalyptus	USA (Cal)	20	45	300	790	19
Eucalyptus	India	17	39	260	678	16
Eucalyptus	Ethiopia	21	48	320	834	19
American sycamore	USA (Ga)	3.7	8	60	160	4
Algae (pond)	USA (Ca)	39	88	580	1,520	36
Tropical rainforest (typical)		18	41	270	710	17
Subtropical deciduous forest (typical)		11	24	160	420	10

iv.   Technical options for using biomass fuels for irrigation pumping

All biomass fuels are ultimately burnt so as to power an appropriate internal or external combustion engine. There is, however, a plethora of options available for preparing, processing or modifying raw biomass for more effective use as a fuel, as shown in Fig. 156. Generally these involve a trade-off between enhancing the properties of the biomass as a fuel on the one hand, and extra cost combined with losses of some of the original material. No particular route can be said to offer advantages over any other, rather there are "horses for courses"; some are better than others in specific applications and situations.

It can be seen from Fig. 156 that there are three primary categories of biomass raw materials:

a.   solid woody ligno-cellulose material;

b.   wet vegetation, residues and wastes;

c.   oil-bearing seeds and resins.

In most cases these need some kind of processing before use; at the very least drying.

4.10.2 The Use of Solid Fuels

As indicated in Fig. 156, solid fuels may be treated in a number of ways and either burned as a solid to power an external combustion engine such as a steam engine, or they may be pyrolised to yield either combustible gas (which may be used for an internal combustion engine), or the volatiles may also be condensed to yield a limited quantity of liquid fuels.

Because, as explained earlier, moisture content of the fuel has a profound effect on its calorific value, it pays to use any waste heat from the system to pre-dry moist fuels.

i.   Direct combustion

Historically a huge range of solid fuel furnaces and boilers existed, but today only a few manufacturers make them for small power systems, although medium sized furnaces and steam plant for use in tropical agro-industrial process plants such as sugar refineries, which are much larger than is appropriate for powering small scale irrigation pumps, are readily available.

Perhaps the most appropriate and simple arrangement for small systems is for fuel to be simply fed by hand into a furnace, containing a boiler to generate steam. Fig. 157 shows as an example, the 2kW experimental Ricardo steam engine developed in the early 1950s. A similar type of unit of 1900 vintage is shown in Fig. 103, for firing a small stirling cycle hot air engine. Furnaces of this kind will typically burn 2-3kg/h of wood per kilowatt of shaft output, with small steam engines.

There are some difficulties in designing a furnace which will handle any fuel; quite different grate arrangements are needed to cope with particulate fuels such as sawdust or rice hulls, as compared with large lumps such as logs or coal. Therefore it is important to use equipment able to accept the proposed fuel. For example, there are furnaces designed especially to handle fuels like sawdust or rice hulls, which would clog up a conventional grate arrangement; in one such type known as a "Kraft Furnace", the furnace and storage hopper are combined so that the outer surface of the mass of sawdust burns and the partially burned gases are drawn through a multitude of small passages into a secondary combustion chamber, where combustion is completed.

ii.   Gasification

The purpose of gasification is to convert some of the energy of an inconvenient solid fuel into a more convenient gaseous fuel. The main advantage of gaseous fuels is that they can generally be used with internal combustion engines and not just with rare steam or Stirling engines; [73].

The first commercially successful i.c. engine powered by a gasifier was built by Lenoir in France in 1860 and ran on coal, [74] and the technique was widely used at the beginning of this century. It was again widely used during the 1939-45 World War, when some 700,000 gasifiers were in use for powering motor vehicles due to shortages of petroleum; [75]. The subsequent cheap oil era led to a great decline in their use, but nevertheless a number of manufacturers in various countries still make gasifier or producer gas units.

The process involves the heating of a solid, carbonaceaous fuel to drive off inflammable volatiles and to produce carbon monoxide (CO) from a reaction between the carbon and the carbon dioxide generated by primary combustion. Moisture in the fuel, plus carbohydrates in the biomass also react with carbon to yield further carbon monoxide plus free hydrogen, and some of the free hydrogen reacts with carbon to produce methane. Any source of heat may be used to gasify biomass fuels, but usually the heating process is by partial combustion; i.e the fuel is burnt to heat itself. The chemical make up of producer gas is typically:

17% CO;   18% H_2;   14% CO_2;   2% CH_4;   49% N_2

and its calorific value will be approximately $5MJ/m^3$. Because of the high proportion of inert nitrogen and carbon dioxide, this is only one eighth of the energy per unit volume of natural gas, such as methane. The calorific value can be enhanced by injecting steam into the gasifier; this yields more hydrogen, but with small systems it is difficult to do this in an optimal manner without actually extinguishing the primary combustion; sometimes there is an advantage therefore if the raw fuel is slightly moist which achieves a similar effect.

Although the calorific value of producer gas is low, the quantity of air required for combustion is also low, so that the thermal value of a stoichiometric mixture, as is required to be induced into an engine for optimum combustion, of producer gas and air is better than might be expected,

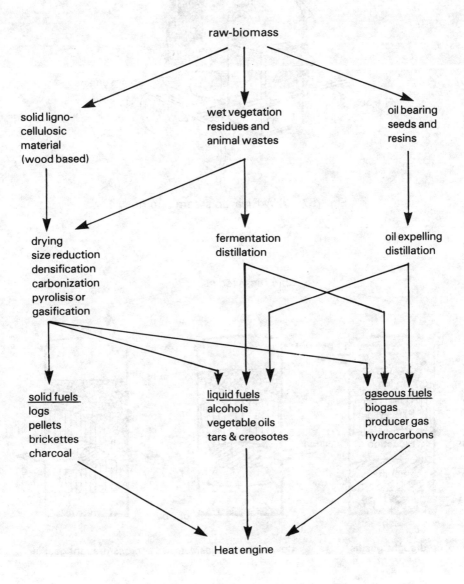

Fig. 156 Routes for processing biomass fuels

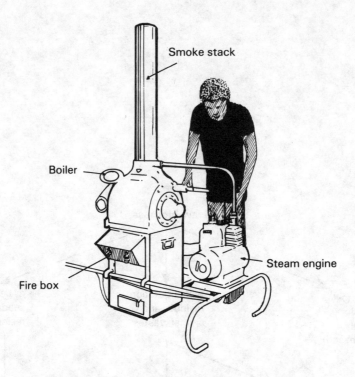

Fig. 157 2kW Ricardo steam engine

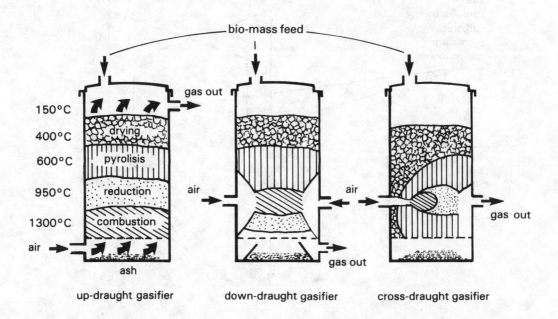

Fig. 158 The three main types of gasifier

as indicated by the following relationships:

Fuel gas	thermal value (MJ/m³)	thermal value of air-fuel mix (MJ/m³)
natural gas (methane)	40	3.5
Producer gas	5	2.5

A producer gas generator is usually a vertically mounted cylinder (see Fig. 158) which is generally loaded with fuel from the top. The fuel falls under gravity to replace burnt and gasified material in the lower fire zone. There are three main types of gasifier, as shown in the figure:

a. up-draught, in which air is introduced at the bottom and the feedstock has the products of gasification passing through it;

b. down-draught, in which the volatiles are drawn downwards through the hottest part of the combustion zone; this has the effect of cracking any tars and complex chemicals and thereby gasifying them more effectively than with an up-draught unit;

c. cross-draught; these have a small intensely hot zone fed from an air nozzle or tuyere.

The first of the above options is simplest, but it produces gas with a lot of carry-over of tar and volatiles which can rapidly damage an internal combustion engine. Therefore down-draught gasifiers are more commonly used for powering i.c. engines as their output is easier to clean, particularly if burning raw biomass fuels containing a lot of volatiles (tar-problems for up-draught units are reduced by using pre-pyrolized fuels like charcoal or coke, but of course much of the original energy content of the feedstock is lost in producing the charcoal or coke). Cross-draught units produce very intense heat in a small area, which results in effective gasification of volatiles and tars, but there are often problems with the nozzle burning out unless it is water cooled (which is a complication); therefore they are less common for use with small engines.

Although down-draught units are preferable for use with i.c. engines, they are less capable of drying the fuel (unlike an up-draught unit the hot gases do not pass through the unburnt fuel), neither can they handle small or particulate fuels so well as these fall through the grate and clog it.

A typical small producer gas irrigation pumping system is illustrated schematically in Fig. 159 (after [76]). Here a down-draught gasifier is used, with wet coke as a primary filter and cotton waste as a secondary filter for the gas.

iii. Producer gas cleaning

Before producer gas can be used in an internal combustion engine, it needs to be effectively cooled and cleaned of impurities including ash, unburnt fuel dust, tar and acidic condensates as otherwise any significant carry-over of these materials will quite rapidly destroy the engine. Obviously the more ash and tar in the original fuel, the more of a problem there is in cleaning the output gas. Therefore fuels having an ash content greater than

5-6% are not recommended for use in producer gas units for i.c. engines. Also, the high performance gasifiers necessary to run small i.c. engines tend to be sensitive to inconsistencies in the fuel, so that regularly sized, low ash fuels are best. Charcoal provides one of the best fuels for gasifiers being almost pure carbon in itself, but coconut shells and maize cobs are both relatively effective gasifier fuels.

The methods used for gas cleaning vary; water or air coolers are generally used to reduce the gas temperature to near ambient conditions, and cyclones, spray scrubbers, filters packed with a wet matrix of wood-wool, steel swarf, coir fibre and other materials have been tried. Ineffective gas cleaning remains the "Achilles Heel" of small gasifier systems, being a major cause of premature engine failure.

iv. Engines for use with gasifiers

After cleaning, the producer gas is mixed with air metered in the appropriate quantity and the resulting mixture can then be induced into the inlet manifold of most standard i.c. engines. Spark ignition engines are capable of running exclusively on producer gas, but diesel engines on the other hand will not fire when run purely on producer gas, and need to be run with at least a small amount of diesel fuel so that the timed injection fires the mixture at the appropriate moment. Therefore they can be run as pilot fuel engines in which diesel is used to start up and continues to be used in quantities normally necessary just for idling, with producer gas making up the main part of the fuel supply. In practice it is possible to run diesels on about one third diesel fuel and two thirds producer gas. The experimental unit illustrated in Fig. 159 [76] is claimed to have actually achieved an average of 88% replacement of diesel, but this may be partly due to unusually careful operation.

The low calorific value of producer gas compared with petroleum fuels generally leads to a marked reduction in power output, often by as much as 30 to 50% below the rated power using petroleum. An approximate idea of the fuel requirements using producer gas compared with conventional diesel operation, is as follows:

Gasifier fuel required to produce 1kWh of shaft power

charcoal	hardwood chips	diesel fuel
1-1.3kg	2-3kg	0.3-0.5kg

Typical producer gas system costs, for units made in Europe or North America (and excluding the engine) are in the region of US $2,500 for a unit capable of fuelling a small engine of about 2kW rating up to around US $4,500 for a unit capable of sustaining a 10kW engine. Less sophisticated units, costing only a few hundred dollars, are manufactured in the Philippines and Brazil. The Indian unit of Fig. 159 [76] has a gasifier costing 8,000 rupees, which is approximately US $800, and is sized for a 3kW engine.

v. Operation and maintenance of gasifiers

A draught is needed to get a gasifier going; this is usually created by starting the engine on a petroleum fuel and then introducing a burning rag or other source to the gasifier fire-zone. Care is needed to ensure that there is no residual gas from the last time the system was run which could explode. The producer gas unit needs to be refuelled before the fuel in the hopper reaches a level of less than about 300mm above the fire zone, or gas production may not be reliable.

Care is needed with gasifiers, since; firstly producer gas is extremely toxic due to the carbon monoxide present and therefore a unit must never be used in enclosed conditions where producer gas could build up; secondly there is a significant risk of explosion and/or fire when opening the unit to refuel it. Opening the gasifier to refuel often causes a small blow-back explosion, but the experienced operator can open the hopper and refuel safely.

The gasifier, and gas cleaning system, must be regularly cleaned out and any leaks must be repaired immediately. Experience with automotive gasifiers during the war suggested that as much as one hour per day on average is needed to clean and prepare a gasifier for operation.

4.10.3 The Use of Liquid Biomass Fuels

There are two main categories of liquid fuels which are relevant to powring small engines for irrigation pumping; alcohols and oils, plus a third (latex or sap) which may come into use in the future.

i.   Alcohols

There are two varieties of alcohol that can be used to run engines; ethanol (ethyl alcohol) and methanol (methyl alcohol). The former is the only type which at present has any prospect to be produced economically from biomass feedstocks (the latter, although once known as "wood alcohol" as this was originally the main route to its production, is usually produced by an industrial process at high pressures and temperatures from natural gas as the process for distilling it from wood is inefficient and unproductive).

Ethanol, which is the type of alcohol found in wines and other drinks, can be produced by the bacteriological fermentation of natural sugars or other carbohydrates such as starches, either from purposely grown fuel crops or from wastes and residues. Starches first require hydrolyzation, usually with acids, to change them into fermentable sugars.

Current activities to produce fuel alcohol focus on the use of sugar cane, maize and cassava [77], processed on a large scale. There is as yet no technically and economically viable small-scale process for fuel-alcohol production, so the use of alcohol by farmers must depend on any national programmes in their countries. They may also be in a position to grow the fuel crops for the programme, so such programmes could be more important for farmers than simply providing an alternative fuel.

There are a number of problems inherent in the large-scale production of fuel alcohol. Obviously the food versus fuel argument, as outlined earlier, is important; secondly, large-scale ethyl alcohol production produces large volumes of "distillery slops", which pose a serious disposal and pollution problem. Finally, the product is only marginally economic at present compared with gasoline and the main justification must generally be import substitution rather than cheaper fuel. Nevertheless a number of alcohol fuel programmes have been initiated. By far the largest is in Brazil, where the goal is to produce 12 billion litres of ethanol per annum by 1985, mainly from sugar cane. The USA also has a major ethanol production programme, based largely on using up maize surpluses, which has the goal of an overall 10% substitution for gasoline by the 1990s. A number of other countries have initiated power alcohol activities, including Thailand, the Philippines, New Zealand, Australia, Kenya, Zambia, Zimbabwe, Nicaragua, Paraguay, and Fiji [77].

R&D is in hand to develop alcohol production processes that could work effectively and economically using woody waste materials and residues, which if successful might create a much more promising future for biomass-based alcohols.

ii. <u>Oils from biomass</u>

There are two types of vegetable oil that show promise as fuel for internal combustion engines, these are expressed oils from seeds and the saps or latex from succulent plants and various trees such as the rubber tree.

Some successes have been reported with running diesel engines on vegetable oils. Tests have been run on seed oils from peanut, rape, soybean, sunflower, coconut, safflower and linseed [78]. Sunflower oil, in particular, shows promise as a fuel for diesel engines. The main problems relate to the much higher viscosity of vegetable oils compared with diesel gasoil; this makes it difficult to start a diesel on vegetable oil, but once warm it will run well on it. Tests have shown that the performance is little effected, but fuel consumption on vegetable oil is slightly higher due to its lower calorific value. A major problem with unmodified vegetable oils has been a tendency for engines to coke up rapidly, leading to reduced power and eventually engine seizure if no corrective action is taken.

Chemical treatment of vegetable oils to turn them into an ethyl or methyl ester has been found to overcome most of these problems and to actually give a better engine performance than with diesel oil, combined with less coking than with diesel [78]. Also, blends of sunflower oil with diesel fuel appear to reduce or eliminate some of the problems experienced with pure sunflower oil.

Large scale processing of vegetable oil can crack the oil in much the same way as crude oil, to produce veg-gasoline as well as veg-diesel. During the Second World War, China developed an industrial batch cracking process for producing motor fuels from vegetable oils, mostly tung oil [79]. The China Vegetable Oil Corporation of Shanghai was able to produce 0.6 tonne of veg-diesel, 250 litres of veg-gasoline and 180 litres of veg-kerosene per tonne of crude vegetable oil.

It is possible to extract vegetable oil "on-farm" on a small scale and to consider using this to reduce diesel fuel requirements, although obviously any vegetable oil needs to be extremely well filtered before it can be used in an engine. A more likely approach would be production on a small-industry basis, in which the extraction unit procured seed from a district for oil production on a more economic scale. Typical seed yields (sunflower) are 700-1800kg/ha. It is possible to express between 0.30-0.43 litres of oil per kg of seed, depending on the technique. Small-scale presses will produce the lower level of yield while large screw presses and solvent extraction are needed to achieve the upper level. This implies that from 210-770 litre/ha can be produced. The development of more efficient on-farm oil seed expellers could make this a potentially viable process in many areas. In fact a combination of efficient cultivation and efficient oil extraction could yield in excess of 1 tonne/ha of vegetable oil. It has been argued that if a mechanized farmer used 10% of his land for sunflower cultivation, he could become energy self-sufficient [79] (although it is not explained by the reference how he overcomes the need to blend his oil with diesel to avoid gumming and coking of the engine).

The energy ratio for the production of vegetable oils as fuel is much more favourable than for alcohol production, and the process is simpler and less capital intensive.

Therefore, the use of oil-seed as a feed-stock to produce diesel fuel certainly looks technically feasible. However the economics remain more doubtful with present diesel fuel prices, since the value of refined vegetable oils on the international market is generally 50-100% above that of diesel fuel, although this price differential has not always applied and may not actually reflect the true cost (as opposed to the price) of vegetable oils.

iii. Latex bearing plants

Some recent investigations have shown a potential diesel fuel can be obtained from the combustible rubbery sap or latex of various trees and succulent plants. Some of these plants actually produce hydrocarbons, similar to but molecularly more complex than petroleum oils. Some promising plants for this purpose are the Euphorbia species [80]. These grow well in semi-arid areas on marginal and barren lands which generally will not support food crops. Professor Calvin of the University of California who has studied this possibility projects a yield of 10 barrels of oil per acre with existing wild species and that this yield could perhaps be doubled through seed selection and genetic improvements to produce a plant developed for oil production [81]. The oil or emulsion which is tapped off has too high a viscosity for immediate use as a diesel fuel, and contains gums and other complex chemicals which would coke the engine prematurely, so it needs to be refined. Although at present this source of fuel remains unproven, it shows considerable promise for the future, particularly if it can be cultivated on lands of little use for food or other crops.

4.10.4 Gas from Biomass: Biogas

At present biogas is the most immediately practicable means for powering a conventional internal combustion engine from biomass. It lends itself to small-scale on-farm use and there is considerable experience with this technique in a number of countries.

Biogas is produced naturally by a process known as anaerobic digestion, the action of bacteria on water-logged organic materials in the absence of air. Biogas occurs naturally as "marsh gas", an inflammable gas which bubbles out of stagnant marshes or bogs. The same process occurs in the digestive system of cattle.

Biogas consists of about 60% methane, a non-toxic and effective fuel gas similar to many forms of natural gas; the remaining 40% is mainly inert carbon dioxide with traces of hydrogen, hydrogen sulphide, etc. Raw biogas has a calorific value of about 23 MJ/m³, which is considerably better than producer gas (see Table 32). The carbon dioxide can be removed by bubbling raw biogas through slaked lime (calcium hydroxide) but this process requires regular replacement of the lime. After this treatment biogas approximates to pure methane with a calorific value of about 40MJ/m³.

Biogas is an attractive fuel for use in i.c. engines since it has no difficult pollutants that can damage them (like producer gas does). Moreover, biogas has good anti-knock properties and can safely be used with high compression ratio spark ignition engines as the sole fuel. When used with diesel engines it needs to be used as a supplementary fuel because a small quantity of diesel fuel needs to be injected to fire the mixture (the injection pump determines the ignition timing). Biogas can be used to reduce diesel fuel requirements by from 50 to 80% with minimal modification to the engine being necessary. To make the best use of biogas requires a spark ignition engine with a compression ratio approaching that of a diesel. Some special biogas engines have been built, which run on 100% biogas more efficiently than with an unconverted gasoline engine [82].

An important further advantage of this process, especially in the context of irrigation pumping, is that the digested sludge makes a good fertilizer, so that unlike the situation where when biomass is totally burnt, it is possible to return much of the original material to the land and thereby improve the soil quality and displace the use of chemical fertilizers. The anaerobic digestion process makes the nitrogen and various other chemicals more accessible to plant growth than the normal aerobic (in air) composting process. Also, unlike artificial fertilizers, the sludge left over from the biogas process contains humus which can improve the soil structure. This process also is useful as a method for treating sewage or disposing of other unpleasant or potentially dangerous organic wastes as well as for producing fertilizer and fuel gas. Anaerobic digestion is a standard sewage treatment process which kills most water-borne pathogens harmful to people and converts the effluent to a relatively innocuous and odourless liquid which can easily be sprayed or poured onto the fields.

Anaerobic digestion is quite widely used for large scale city sewage plants, but it is also increasingly being applied on farms. The first reasonably widespread farm use was in France during the Second World War, when farmers built concrete digesters to produce methane to replace petroleum fuels which were unobtainable for them at that time. More recently efforts have been made to popularise the use of biogas in Asia, mainly in China, but also in India, Nepal and some of the SE Asian countries. Commercial farm biogas units have also gone into production in various countries, including the USA, UK, Australia and Kenya as well as the main users of the technology, China and India.

Although the widespread use of biogas only started in China in the early 1970s, within ten years some seven million biogas units had been installed [83], with the majority being in Sichuan Province. The technology has been less successful at spreading in India, although some 80,000 digesters are believed to be in regular use there. Experience in India has been that the larger biogas units used by richer farmers and by institutions have

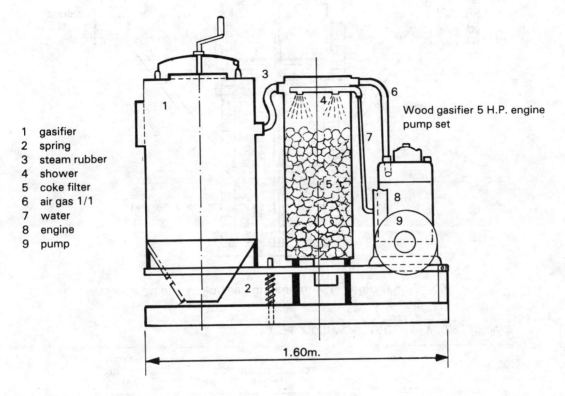

Fig. 159 Small producer gas irrigation pumping system (ref. Damour [76])

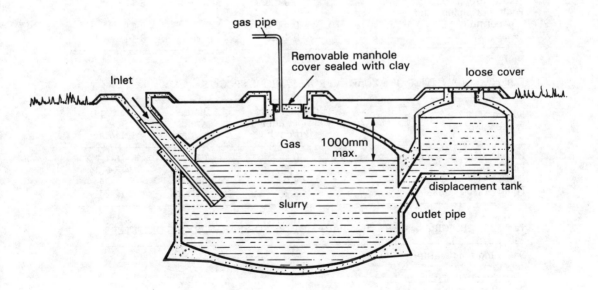

Fig. 160 Fixed dome biogas digester (China)

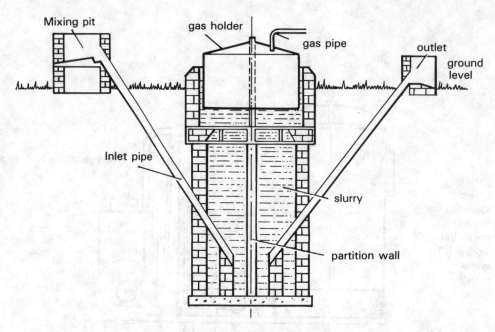

Fig. 161 Biogas digester with floating gas holder and no water seal (India)

Table 34 BIOGAS YIELD FROM VARIOUS FEEDSTOCKS

Feedstock	Gas yield per unit mass of feedstock (m³/kg)	energy yield (MJ/kg)
Sewage sludge	0.3-0.7	6-17
Pig dung	0.4-0.5	8-11
Cattle dung	0.1-0.3	2- 6
Poultry droppings	0.3-0.5	6-11
Poultry droppings and paper pulp	0.4-0.5	8-11
Grass	0.4-0.6	8-14

Table 35 QUANTITIES OF EXCRETA FROM VARIOUS SPECIES

Source of waste	Volatile solids yield per animal (kg/day)	Biogas yield (m³/day)	Energy yield (MJ/day)
Humans (inc. cooking wastes)	0.1	.03-.07	.6-1.7
Pigs	0.6	.24-.30	4.8-6.6
Cattle	4.0	.40-1.2	8-24
Poultry (x100 birds)	2.2	.07-1.1	13-24

Table 36 PRINCIPAL OPERATING PARAMETERS FOR FARM BIOGAS DIGESTERS

Operating temperature	30-35°C
Retention time	20-40 days
Loading rate (volatile solids)	2-3 kg/m³ per day
Operating moisture content	85-95%
Specific gas production	0.1-0.7 m³/kg per day
Feedstock carbon/nitrogen ratio	20-30

generally been more successful than smaller "family-sized" units.

Figs. 160 and 161 illustrate the two main types of small-scale biogas digester, developed originally in China and in India respectively. The Chinese type of digester consists of a concrete-lined pit with a concrete dome, entirely below ground. It is completely filled with slurry, and once gas begins to form, it collects under the dome and forces the level of the slurry down by up to about 1m. The gas pressure is consequently variable depending on the volume of gas stored, but by using a simple manometer on the gas line it is possible to measure the gas pressure and thereby gain an accurate indication of the amount of gas available. The Indian type of digester (Fig. 161) is more expensive to construct because it has a steel gas holder, on the other hand it is less likely to leak than the Chinese design which requires high quality internal plastering to avoid porosity and hence gas leaks. With the Indian design, gas collects under the steel gas holder, which rises as it fills with gas. The height of the gas holder out of the pit indicates how much gas is available, and the pressure is constant.

i.   The biogas production process

The biogas process requires an input material provided as a liquid slurry with around 5-10% solids. It is important to use materials which breakdown readily; highly fibrous materials like wood and straw are not easily diegested by the bacteria, but softer feedstocks like dung and leaves react well to the process. Also some feedstocks are more productive than others as indicated by Table 34, and some producers of feedstock are more productive than others as indicated by Table 35.

Table 36 (after Meynell [84]), gives the principal operating parameters of typical continuous biogas digesters (it is also possible to run batch digesters in which each digester is loaded, completes its cycle and is then unloaded, but this needs several units, probably three, to ensure gas is always available). For optimum performance the internal temperature of the digester needs to be in the mid-30s centigrade and certainly over 25°C, moreover temperature conditions need to be as steady as possible. The digestion process generates a small amount of heat, but in cooler climates or seasons the unit needs to be well insulated and may need heating when cold spells occur. The average retention time for solids for the complete process is normally 20 to 40 days. With continuously operating (as opposed to batch) digesters, the actual digester size has to be equal to the design retention time in days multiplied by the daily input rate; i.e. with a 30 day retention time and $1m^3$/day of input, the digester volume needs to be $30m^3$. The longer the retention time and the warmer the digester, the more complete the process and the more energy per kg of volatile solids is obtained, however the larger and more expensive the digester needs to be. Hence the sizing and retention time are usually a compromise between keeping costs reasonable and obtaining complete digestion. The loading rate and the moisture content are related; the slurry needs to be kept to around 85 to 95% liquid. Too thin a slurry takes up more volume and needs a bigger digester than necessary, while too thick a slurry limits mixing and tends to solidify and clog up the unit. Another important criterion is the carbon/nitrogen ratio; for efficient digestion the process requires between 20 and 30 parts of carbon to be present per part of nitrogen. Certain carbon-rich materials like leaves or grass benefit, therefore, from being mixed with nitrogen-rich substances such as urine or poultry droppings. Alternatively ammonia or other nitrogen rich artificial chemicals may be added to a digester running on mainly vegetation to obtain a batter ratio and help the process. Finally, the output to be expected will be in the order of 0.1-0.7m³ per kg of volatile solids input per day.

Because biogas digesters have the capability of storing at least a 12 hour supply of gas, an engine can be used that draws gas at quite a high rate. In fact the size of engine is not critical since it is only the number of hours it will run that are governed by the digester gas capacity. Transporting biogas is technically difficult. In China it is quite often piped several hundred metres through plastic tubes. Unlike propane or butane, it is not possible to compress biogas into a liquid at normal temperatures and the only way to transport it as a gas are either in high pressure cylinders, which of course require a high pressure compressor to charge them, or in a plastic bag. Fig. 162 shows how small two-wheel tractors in China are powered from a bag of biogas carried on an overhead rack. The unit in the figure is towing a trailer tank full of biogas digester sludge and it also carries a pump driven off the engine for pumping the sludge onto the field via a spraying nozzle. An interesting option for irrigation by biogas power is to combine the digester sludge and the irrigation water in order to perform three functions simultaneously; i.e. irrigation, the application of fertilizer and waste disposal.

ii.   Sizing example

Biogas typically has a calorific value as a fuel for running small engines of about 6.4kWh/m³, so it is quite straightforward to estimate the daily volume of biogas needed to perform a given pumping requirement. A worked example of how to do this is given in Table 37, which indicates how a 3ha small-holding could be irrigated using biogas generated from the wastes from 20-30 pigs, 5-10 cattle, 500-700 poultry or a community of 80-200 people. The production rate of biogas can be enhanced by mixing vegetation with the animal wastes, although extra nitrogen, which could be in the form of urine, may need to be introduced to balance the excess carbon present in the vegetable wastes.

The example therefore shows that this process needs significant inputs of waste material to yield even quite modest amounts of pumped water. Therefore, looked at just in energy terms, the economics tend to be at best marginal in comparison with petroleum fuels, however when the fertilizer value plus the waste disposal benefits are factored in, the process frequently comes out as being economically worthwhile.

It is difficult to generalize on the economics of biogas since many factors that are locality-specific are involved. However there is no doubt that the process offers significant economies of scale. For example a survey [85] of biogas units in India found a payback period, using a 10% discount rate, of 23 years for a 1.7m³ (60 cu.ft.) plant which improved to 7, 4 and 3 years respectively with 2.8, 5.7 and 8.6 m³ units (100, 200 and 300cu.ft). The sizes of plants needed to run small engines are much bigger than this and are therefore likely to be more cost-effective.

The country which has made by far the most use of biogas production in agriculture is China, where, particularly in Sichuan Province, there are

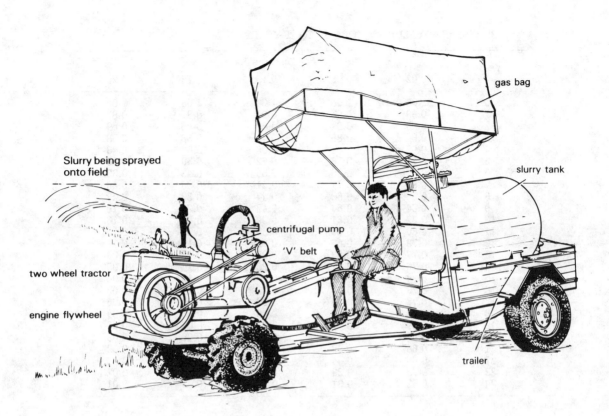

Fig. 162 Chinese two-wheel tractor running on biogas and being used to pump digester slurry on to the field

Table 37 SIZING EXAMPLE TO RUN AN IRRIGATION PUMP ON BIOGAS

Requirement: 8mm of water per day pumped through a head of 6m (i.e. 240m³/day)

Engine: s.i. engine assumed 10% efficient (fuel to hydraulic power)

Biogas: calorific value assumed at 6.4kWh/m³

The energy requirement to lift 240m³/day of water through 6m is:

$$E_{hyd} = \frac{QH}{367} = \frac{240 \times 6}{367} = 3.92 \text{kWh/day (see Section 2.1.8)}$$

Assuming a system efficiency of 10%, the fuel energy requirement is:

$$E_{fuel} = \frac{3.92}{0.1} = 39.2 \text{kWh/day}$$

Hence the daily biogas requirement will be:

$$Q_{biogas} = \frac{39.2}{6.4} = 6.125 \text{ m}^3/\text{day}$$

This requires a biogas digester of about 5-10m³ capacity, loaded with from 10-60kg/day of input material (volatile solids), which can in turn be obtained from, for example:

    20-30 pigs   5-10 cattle   500-700 poultry   80-200 people

The addition of vegetable wastes, providing it did not unduly upset the carbon/nitrogen ratio, could allow the same volume to be produced from possibly two thirds to three quarters the number of livestock or people.

Table 38 PRESENT VALUE FACTORS UP TO 25 YEARS

Year	Discount Rate				
	0.02	0.05	0.10	0.15	0.20
0	1.00	1.00	1.00	1.00	1.00
1	0.98	0.95	0.91	0.87	0.83
2	0.96	0.91	0.83	0.76	0.69
3	0.94	0.86	0.75	0.66	0.58
4	0.92	0.82	0.68	0.57	0.48
5	0.91	0.78	0.62	0.50	0.40
6	0.89	0.75	0.56	0.43	0.33
7	0.87	0.71	0.51	0.38	0.28
8	0.85	0.68	0.47	0.33	0.23
9	0.84	0.64	0.42	0.28	0.19
10	0.82	0.61	0.39	0.25	0.16
11	0.80	0.58	0.35	0.21	0.13
12	0.79	0.56	0.32	0.19	0.11
13	0.77	0.53	0.29	0.16	0.09
14	0.76	0.51	0.26	0.14	0.08
15	0.74	0.48	0.24	0.12	0.06
16	0.73	0.46	0.22	0.11	0.05
17	0.71	0.44	0.20	0.09	0.05
18	0.70	0.42	0.18	0.08	0.04
19	0.69	0.40	0.16	0.07	0.03
20	0.67	0.38	0.15	0.06	0.03
21	0.66	0.36	0.14	0.05	0.02
22	0.65	0.34	0.12	0.05	0.02
23	0.63	0.33	0.11	0.04	0.02
24	0.62	0.31	0.10	0.03	0.01

Table 39 ANNUALIZATION FACTORS UP TO 25 YEARS

Year	Discount Rate				
	0.02	0.05	0.10	0.15	0.20
0	1.00	1.00	1.00	1.00	1.00
1	1.02	1.05	1.10	1.15	1.20
2	0.52	0.54	0.58	0.62	0.65
3	0.35	0.37	0.40	0.44	0.47
4	0.26	0.28	0.32	0.35	0.39
5	0.21	0.23	0.26	0.30	0.33
6	0.18	0.20	0.23	0.26	0.30
7	0.15	0.17	0.21	0.24	0.28
8	0.14	0.15	0.19	0.22	0.26
9	0.12	0.14	0.17	0.21	0.25
10	0.11	0.13	0.16	0.20	0.24
11	0.10	0.12	0.15	0.19	0.23
12	0.09	0.11	0.15	0.18	0.23
13	0.09	0.11	0.14	0.18	0.22
14	0.08	0.10	0.14	0.17	0.22
15	0.08	0.10	0.13	0.17	0.21
16	0.07	0.09	0.13	0.17	0.21
17	0.07	0.09	0.12	0.17	0.21
18	0.07	0.09	0.12	0.16	0.21
19	0.06	0.08	0.12	0.16	0.21
20	0.06	0.08	0.12	0.16	0.21
21	0.06	0.08	0.12	0.16	0.20
22	0.06	0.08	0.11	0.16	0.20
24	0.05	0.07	0.11	0.16	0.20
24	0.05	0.07	0.11	0.16	0.20

several million working biogas units. This development took place almost entirely within the last 10-15 years. Various studies (eg. [86]) have indicated that the value of the fertilizer output usually surpasses the value of the energy produced by the process in China. The waste disposal and sanitation aspects of the process are also important justifications for its use.

5. THE CHOICE OF PUMPING SYSTEMS

5.1 FINANCIAL AND ECONOMIC CONSIDERATIONS

5.1.1 Criteria for Cost Comparison

The ultimate criterion for choosing an irrigation pumping system is to obtain the most "cost-effective" system; this does not necessarily mean the "cheapest" system, since low first cost often results in high running costs. To arrive at a realistic assessment of the true cost-effectiveness is not easy, particularly as many of the parameters required for such an analysis will often be uncertain or variable, and many "costs" and "benefits" do not readily lend themselves to financial quantification at all (eg. reliability, availability of spare parts or maintenance skills, ease of use and vulnerability to theft are some factors which have cost implications, but which are difficult to factor in to any cost analysis). Nevertheless, the objective of most methods of financial or economic analysis is to arrive at a figure for the true "life-cycle costs" of a system (i.e. the total costs of everything relating to a system over its entire useful life) which are often compared with the "life-cycle benefits", which are the total benefits generated by the system in its lifetime. Since the length of life to be expected for different options will vary, it is necessary to find a technique for reducing the life cycle costs and benefits to those over, say, one year or to find some other method of presentation which allows ready comparison between options. Some methods for achieving this are discussed in more detail later.

While a financial or economic appraisal of options generally represents the primary criterion for selection, this should not necessarily be used as the sole method for ranking. Obviously any clearly uneconomic type of system is to be rejected, but, generally, there will be a number of possible options which in purely economic or financial terms show similar costs. The final selection from such a "short-list" generally needs to be based on technical or operational considerations. It is a mistake to use economic or financial analysis as the sole arbiter for choosing between options.

It is important to distinguish between economic and financial assessment of technologies. An economic assessment seeks to look at "absolute" costs and benefits and therefore considers costs and benefits as they would be if unaffected by taxes, subsidies or other local influences; the object is to arrive at a valuation of the technology in pure terms, excluding any local financial conditions. The value of an economic assessment is more for policy makers and those who need to compare technologies. The farmer, on the other hand, should do a financial assessment which takes account of conditions within his local economy, such as subsidies and taxes and the local market price of the final product harvested as a result of irrigation. The economist and the farmer may therefore come to different conclusions as to what is "cost-effective".

There are two main cost factors relating to any system; its first cost (or capital cost) and its recurrent costs or operation and maintenance costs (O&M costs). So far as the first cost is concerned, the over-riding consideration from the individual farmer's point of view is whether it is affordable as a cash payment and if not, whether he can obtain finance on acceptable terms. The institutional user will no doubt have access to finance and will be more concerned to ensure that an adequate return on the investment will be obtained. The O&M costs can vary considerably both within and between technologies and will no doubt increase with time due to inflation. Generally speaking, all options for pumping water represent a trade-off between capital costs and recurrent costs; low first cost systems usually have high recurrent costs and <u>vice-versa</u>.

Clearly, an investment in extra costs at the procurement stage may save recurrent costs throughout the life of the system, but it is difficult to compare a low cost engine having high running costs with a high cost solar pump having virtually zero running costs. Therefore, a major problem which may often inhibit the choice of high capital cost systems is the difficulty involved in comparing the cost of such an investment with the savings it could produce for many years into the future.

5.1.2 Calculation of Costs and Benefits

In all cases it becomes necessary to use some method to compare the notional value of money in the future with its value today, so that such things as the trade-offs between spending more on the first cost in order to reduce the recurrent costs can be properly assessed. All the methods used rely on what is called Discounted Cash Flow (DCF). The general principle behind this is that money available in the future is worth less than if it were available now; this is not necessarily because of inflation, but because it is assumed that money available now can be invested and will therefore be worth more in the future than the original sum. For example, investing $100 now at 10% will yield $110 in one year's time; therefore $100 today is notionally worth 10% more than the same sum if made available in one year's time.

Discounted Cash Flow therefore takes all future payments and receipts relating to a predictable cash flow, and discounts them to their present value using an appropriate interest rate, or discount rate. Going back to the previous example, $110 received in one year's time is said, therefore, to have a "present value" (or PV) of $100 if discounted to the present at a 10% discount rate.

When capital costs are annualized, then first costs can be properly equated with recurrent costs as far into the future as is desired. The way this is done is to note that it is unimportant to an investor whether he has, say, $100 today which he plans to leave in a deposit account yielding 10% interest, or whether he simply has $110 promised to him in one year's time. It follows from this that, for example, a payment of "C_r" in different numbers of years' time will be discounted back to the present, assuming a discount rate of "d" in the following way:

Years:-	1	2	3	n
PV:-	$\dfrac{C_r}{(1+d)}$	$\dfrac{C_r}{(1+d)^2}$	$\dfrac{C_r}{(1+d)^3}$	$\dfrac{C_r}{(1+d)^n}$

Each of these factors gives the PV of a payment C_r in 1,2,3,...,n years time.

Table 38 gives calculated values for these factors for discount rates of 2, 5, 10, 15 and 20% for all years from 1 to 25. To find the Present Value of, say $1,000 to be paid in 10 years time at a discount rate of 10%, the relevant PV factor is looked up from the table (0.39 in the example) and multiplied by the sum of money in question to give, in this example 0.39 x $1,000 = $390. If it can be anticipated that $1,000 needs to be paid, say, every 5 years (for example to replace an engine), then the PV factors for years 0, 4, 9, 14, 19 and 24 at the relevant discount rate are looked up, added together and multiplied by the sum in question; in this case the calculation for a 10% discount rate and 25 year period would be:

$$(1 + 0.68 + 0.42 + 0.26 + 0.16 + 0.10) \times \$1,000 = \$2,620$$

A complete cash flow over some predetermined future period can be planned and when multiplied by the relevant PV factors for each year can be reduced to a complete Present Value.

Another effect to be considered is that of inflation. The real purchasing power of money tends to decline, so that for example a litre of diesel fuel in one year's time may be anticipated to cost say 5% more than it does today. So if we wish to compare the cost over an extended period, we need to allow for the likely increase in cost of fuel. Therefore if we expect to spend $100 this year, we must plan to spend $105 next year and so on. This results in a need to increase the actual amount of currency required to achieve a given real value. This can be anticipated by assuming an inflation rate "i", and as a result the progression of PV values discounted back to the present becomes:

Years:-	1	2	3	n
PV:	$Cr\left(\frac{1+i}{1+d}\right)$	$Cr\left(\frac{1+i}{1+d}\right)^2$	$Cr\left(\frac{1+i}{1+d}\right)^3$	$Cr\left(\frac{1+i}{1+d}\right)^n$

Whether or not inflation is included, the actual life cycle cost of a system will be the sum of the present values of all its initial costs, plus all its future costs; i.e. the life cycle cost is obtained by taking the capital cost and adding on the anticipated discounted present values of the future recurrent costs.

For most purposes of comparison it is easiest and quite sufficient to ignore inflation and work entirely in "present day money", especially as future inflation is not readily predictable over long periods of time. Therefore Table 38 uses zero inflation rate and no attempt has been made to provide a similar set of tables to cater for various rates of inflation.

A regular payment or benefit "C_a" which recurs every year can be simultaneously inflated at a rate "i" and discounted at "d". To avoid calculating each year separately as above and having to add them together, the series can be summed by a general formula for a whole string of regular payments, as follows:

$$\text{PV to year "n"} = C_a \left(\frac{1+i}{1+d}\right) \frac{\left(\frac{1+i}{1+d}\right)^n - 1}{\left(\frac{1+i}{1+d}\right) - 1}$$

clearly, if $i = d$, then $PV = C_a$.

Having produced a Present Value for some cash flow stretching into the future, it is then important to be able to reduce this to an annual equivalent sum in present day money; in other words, how much money do you have to save each year to finance the cash flow? The annual equivalent "C_r" of a capital sum "C" is similar to paying an annuity and can be calculated as follows:

$$C_r = C \frac{d(1+d)^n}{(1+d)^n - 1}$$

Table 39 gives calculated annualization factors for various discount rates and periods; for example, the annualized payment due to pay off a sum of $1,000 (in present day value money) over ten years at a discount rate of 10% can be calculated by taking the relevant factor (0.16 in this case) and multiplying it by the capital sum to arrive at $160/yr for this example. Therefore is you borrow $1,000 today under those financial conditions, the repayments, or the cost of financing the loan will total $160 per year. Even if you pay $1,000 cash for some item with a 10 year estimated life, the "opportunity cost" resulting from not having the cash available to earn interest in a bank will still be $160 in the case considered, so it is correct to say that it costs $160 per $1,000 tied up over 10 years at a 10% discount rate. Table 39 can readily be used to calculate annualized values of other capital sums over a range of periods up to 25 years for the same discount rates as for the PV factors of Table 38.

Table 39 can also be used in reverse to calculate the Present Value of a regular cash flow; using the same example, the Present Value of $160/yr over a 10 year period at a 10% disount rate is exactly $1,000. The method to use Table 39 to obtain the Present Value of a cash flow is to look up the factor relating to the period and discount rate required, and to divide the annual payment by the factor. For example, the Present Value of $80/yr over a period of 25 years at a discount rate of 15% is the factor 0.03 divided into $80, i.e.:

$$\$80 / 0.03 = \$2,667$$

This is a lot easier than taking the PV factor for every year from year 0 to 24 from Table 38, adding them and multiplying by 80, which should give the same result.

Having explained the methods of equating future costs and payments to the present, it remains to use these techniques to arrive at a means for comparing the relative economic merits of different systems. There are in fact four commonly used techniques for making economic appraisals:

a. <u>Life-cycle-costs and Unit-output costs</u>

The total present value of all costs for various systems sized to do a certain job can be compared (i.e. it is assumed that similar benefits will be achieved regardless of the choice of technology, so the problem is simplified to finding the least-cost solution). This can be confusing if different options have significantly different life expectations (in years), so the lifecycle costs are usually annualized as described above to give a more general means for comparison which includes an allowance for the expectation of life of each system. The annualized life cycle costs can then readily be converted to unit output costs; i.e. the cost of unit quantity of water through a given head, as described above. Where different heads may be involved, the "unit hydraulic output costs" can also be determined by taking the cost of the head-flow product, but care is needed to ensure that you are still comparing like with like as some systems' costs will change significantly as a function of head. Life-cycle costs are purely a criterion for comparison as they do not indicate whether a specific water pumping system is actually economically viable (for example whether the value of additional crops gained from pumping irrigation water may actually exceed the cost of pumping the water).

b. Net Present Value (NPV) or Net Present Worth

This is be obtained by determining cash flows for the benefits expected as well as for the costs, using exactly the same methods as described above. The benefits will generally be the marginal income gained from irrigation and will be given a positive value and the costs, calculated as before will be made negative. To obtain the Net Present Value the present values of the total costs and benefits cash flow are added together. To be worthwhile a positive Net Present Value is required, and the more positive the better; i.e. the summed present values of the benefits must exceed those of the costs. The answer obtained will differ depending on the choice of discount rate, and the period considered for analysis, so this is not therefore purely dependent on choice of technology.

c. Benefit/Cost Ratio

A variation on the concept of Net Present Value is to calculate the total life-cycle benefits and the total life-cycle costs and then to divide the former by the latter to obtain the Benefit/Cost Ratio. If this ratio is greater than one, then the benefits exceed the costs and the option is worthwhile. The same criticisms apply to this as to the Net Present Value approach.

d. Internal Rate of Return

Internal Rate of Return is difficult to calculate, but provides a criterion for comparison independent of any assumptions on discount rates or inflation. It is therefore a purer method for comparing technologies. The Internal Rate of Return can be defined as the discount rate which will give a Net Present Value of Zero (or a Benefit/Cost ratio of 1); i.e. it is the discount rate which exactly makes the benefits equal the costs. To calculate Internal Rate of Return requires finding the discount rate to achieve a Net Present

Value of zero; this is usually determined by trial and error, by recalculating the Net Present Value for different discount rates until the correct result is achieved. It is tedious to do this manually, but various standard micro-computer spread sheets are widely available which make it a relatively easy task. The advantage of the Internal Rate of Return as a selection criterion is that it is, in effect, the discount rate at which an option just breaks-even. If the Internal Rate of Return is higher than the actual discount rate, then the option can be said to be economically worthwhile. Obviously, the higher the Internal Rate of Return of an option, the more attractive it is as an investment, since it basically says whether you do better to leave your money in the bank or invest it in an irrigation pumping system, or whatever.

The choice of discount rate used for analysis effectively reflects the analyst's view of the future value of money; a high discount rate implies that money available now is much more useful than money available in the future, while a low discount rate is more appropriate when longer term considerations lie behind an investment decision. The implications of this are that low discount rates favour the use of high-first-cost low-recurrent-cost systems, which cost a lot now in order to save money in the future, while high discount rates make low-first-cost high-recurrent-cost systems look good be making the high future costs less important.

The length of period selected for analysis (n) can affect the answer, so it is normal to use a long period in order to minimize this effect; this will normally need to be 15 to 20 years or more because at normal discount rates, costs more than about 20 years into the future become discounted to such small levels that they cease to affect the results very much. Obviously, short-life equipment will need replacement during the period under analysis, and this is taken account of by adding the discounted value of the capital costs of future replacements to the life cycle costs.

i. Shadow pricing

Economists recognize that there are so-called "opportunity costs" associated with cash transactions; for example, although for most analytical purposes the exchange rate of a local currency will be taken at the official rate, in practice this does not always reflect its real value in terms of purchasing power. The opportunity cost of using foreign currency is therefore often higher than the exchange rate would suggest. It is therefore legitimate in a comparative analysis to penalize options involving a lot of foreign exchange to a greater extent than would apply simply by using the prevailing exchange rate. The normal method of doing this is to multiply the actual financial cost by a so-called "shadow price factor"; where there is a shortage of a commodity (eg. frequently, diesel fuel) it will have a shadow price factor greater than unity, conversely where there is a surplus (such as being able to use unskilled labour), then a shadow price factor of less than one may be applied to unskilled labour wages. For example the shadow price of diesel fuel may in some rural areas of developing countries be as much as four times the real price, while the ready availability of unskilled labour may allow it in reality to be costed at as little as 70% of its real wage level for economic comparisons. Tables of shadow prices specific to different countries, and even to regions of countries have been developed, but the concept of shadow pricing is complex and is probably best left to economists.

However the principles of shadow pricing should at least be borne in mind, at least since items with a high opportunity cost may go into regular short-supply and cause operational problems.

ii. **A procedure for a cost appraisal of an irrigation pumping system**

Fig. 163 outlines a method that can be used to compare the costs of alternative water lifting techniques. This step-by-step approach is based on life-cycle-costing of the whole system. It takes into account all the identifiable costs, but ignores benefits gained by the users of water.

An integrated approach which considers the system as a whole, from the water source, to the point of application on the field is recommended; i.e. including the water source costs (such as well digging) and water distribution costs (such as digging ditches or purchasing pipes or sprinklers).

iii. **Example: comparison of Engine, Wind and Solar Pumps**

A simple worked example is included as Table 40. The first step is to determine the hydraulic energy requirements. Suppose we wish to irrigate 0.5ha to a depth of 10mm while pumping through a static head of 4m in the month of maximum water demand. Reference to Fig. 13 in Chapter 2 indicates that this requires a nett daily hydraulic energy output of 0.545kWh (factoring down by ten from the scale used in the figure). Alternatively, the following relationship, also explained in Chapter 2, may be used:

$$E_{hyd} = \frac{Q H}{367} \text{ kWh/day}$$

Allowance must now be made for distribution losses; for convenience it is assumed that all three systems being compared involve the same distribution efficiency of 60%. Then the gross hydraulic energy requirement is (0.545/0.6) = 0.91kWh/day for all three. It should be noted that in reality, different distribution efficiencies might occur with different types of system, which would result in different energy demands.

The next step is to determine the design month; this is generally the month of maximum average water demand if the power supply is uneffected by climatic conditions, (eg. for engines), but where the energy resource is the wind or the sun, it becomes necessary to compare the energy demand with the energy availability and the design month will be the month when the ratio of energy demand to energy availability is highest. Supposing in this case that the design month does coincide with the month of maximum water demand in all three examples, then using the assumptions in Table 40, the bases for which are discussed in more detail in the relevant earlier chapters, we arrive at the sizing for the systems. The engine in this case will be of the smallest practical size, but the wind and solar pumps need to be suitably sized.

The next step is to estimate the installed capital cost of the system, generally by obtaining quotations for appropriately sized equipment. Some "typical" values, valid at the time of writing, have been used; namely $200/m² of rotor for a windpump and $15/Wp for a solar pump. The product of size and cost factor gives the installed capital cost, which for simplicity is assumed to include the water source and distribution system in all three cases. Some storage facility is likely to be needed in all cases; a secure lock-up holding two 200 litre oil drums is assumed for the diesel while a low cost compacted

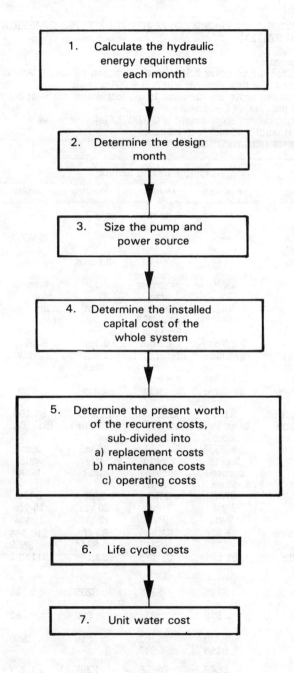

Fig. 163 Step-by-step procedure for a cost appraisal of a water pumping system

Table 40 ANALYSIS OF UNIT WATER COSTS FOR FOUR TYPES OF IRRIGATION PUMPING SYSTEM

Operational requirement:
    10mm of water lifted 4m to cover 0.5ha (peak irrigation demand). Annual requirement averages 67% of peak for five months, which is 5,094m³/year.
    Assumed water source costs are identical in all four cases, for simplicity, although in practice this may not always be correct.
    Peak daily hydraulic energy requirement is 0.92kWh/day
    Total irrigation demand: 5,094m³/yr in all cases.
    Financial parameters (all cases) $D = 10\%$ $N = 25$ yrs $i = 0$

Parameter	Gasoline	Diesel	Wind	Solar	Notes
Price of fuel delivered to field	50¢/litre	40¢/litre	—	—	
Critical month mean irradiation	—	—	—	5.8kWh/m²	
Critical month mean windspeed	—	—	3.5m/s	—	
Sizing assumption	1kW 3% effic.	minimum size available i.e. 2kW & 8% effic.	0.1V³ W/m²	35% mean motor-pump effic.	
Requirement to produce peak daily water output	3.1 l/day gasoline	1.2 l/day diesel fuel	9.7m² rotor area	540Wp array	
Requirement to produce mean daily water output	2.1 l/day gasoline	0.8 l/day diesel fuel	ditto	ditto	
Capital cost assumption (total power system and pump)	$330 (engine & pump)	$1,500 (engine & pump)	$1,940 ($200/m²)	$8,100 ($15/Wp)	
Storage tank capacity	2 x 200 l fuel in secured shed	2 x 200 l fuel in secured shed	40 m³ water tank	30m³ water tank	Cost includes estimated average fuel inventory cost as well as shed cost. Hence gasoline storage costs more on average.
Cost of storage	$280	$250	$600	$450	
Life of system	3 yrs	7 yrs	20 yrs	15 yrs	
Life of storage	15 yrs	15 yrs	15 yrs	15 yrs	
Lifecycle system costs	$1,224	$2,865	$2,405	$10,444	
Lifecycle storage costs	$347	$310	$744	$558	
Total lifecycle capital costs (at present value)	$1,571	$3,175	$3,149	$11,002	
Annualised system costs	$135	$315	$265	$1,166	
Annualised storage costs	$38	$34	$82	$61	
Annual O & M costs	$220	$200	$50	$50	
Annual fuel costs	$156	$49	—	—	
Total annual cost:	$549	$598	$397	$1,277	
Average unit cost of water	10.8¢/m³	11.7¢/m³	7.5¢/m³	25.1¢/m³	

soil bund having a cement lining is assumed for water storage for the wind and the solar systems, holding 40 and 30m³ respectively. In the case of petroleum fuelled systems, the cost of storage includes the notional investment in stored fuel (assuming on average that the storage is at 50% capacity and amortizing a continuous investment in 100 litres of fuel).

The actual useful life of the systems and storages is assumed, as indicated in the table, as are financial parameters for the discount rate and the period for analysis. Hence the life-cycle costs may be determined by working out the present values of the first system and all subsequent replacement ones (using factors from Table 38) and adding them all together. Table 40 shows the system and storage life-cycle costs separately, but they could also be combined. In order to arrive at a comparable annualized cost for the capital investment in each system, the factors of Table 39 are used to convert the life-cycle cost; in this case a 25 year period and 10% discount rate gives a factor of 0.11, which the life-cycle cost needs to be multiplied by. The storage system was dealt with similarly.

The combined system and storage annualized costs represent the annual investment or "finance" costs. Different systems also have recurrent costs consisting of O&M (operational and maintenance) costs, and sometimes fuel costs. When the finance, O&M and fuel costs are added, we obtain the gross annual cost of owning and operating each system.

Where an identical useful output is to be produced, then the gross annual cost is sufficient for ranking purposes. In reality, however, different cropping strategies may apply for different irrigation systems, resulting in different crop irrigation water demands and different benefits (in terms of the market values of the crops). Therefore it is useful to divide the gross annual cost by the gross annual irrigation water demand to arrive at an average unit cost for water from each option.

In this example, the windpump comes out marginally better than the petrol engine but the decision probably ought to be made between them on other than economic grounds as there is so little to choose between them in simple unit cost terms. In this example, the solar pump does not seem economically competitive. It must be stressed that this is but one simple example which should not be blindly used to draw any conclusions on the relative merits of engine, wind and solar pumps generally. Even varying totally non-technology dependent parameters such as the discount rate, the period of analysis, the water demand or the head could significantly change the results and rankings obtained, and so could changing the technical performance and/or cost parameters, which would have an even more profound effect.

5.1.3 relative economics of different options

A procedure similar to that just described has been followed to analyse a representative selection of the types of water lifting systems described earlier in this paper.

Most studies attempting this kind of analysis use a single assumption for each and every parameter and compound these to arrive at a single answer, as in the example just given, often presented as a single curve on a graph for each option. The trouble with this approach is that errors are compounded and may not cancel out, so the result could be very misleading. In an attempt to minimize this problem, the approach in this case has been to choose a low and a high parameter at each and every decision point; i.e. a plausible pessimistic and a plausible optimistic one. Two sets of calculations are then completed for each technology, to produce a pessimistic and an optimistic result, which when graphed give two curves. It is then reasonble to assume that the real result is likely to lie between the two curves and the results are therefore presented as a broad band rather than a thin line. Therefore, where the broad band for one technology lies wholly above or below another it is reasonable to assume the lower one is the cheaper option, but obviously many options overlap considerably and in such situations other considerations than water cost should dictate the decision.

Table 41 lists all the systems considered and gives the principle assumptions used for calculating the cash flows. The capital cost assumptions are intended to include the entire system as defined in the previous section; i.e. not just a prime-mover and pump, but all the necessary accessories that are necessary and appropriate for each type of technology and scale of operation.

Because of the large numbers of options to be analysed, calculations were carried out on a computer and the results printed out graphically as the cost of hydraulic energy versus the peak daily demand for hydraulic energy (see Figs. 164 to 168).

To eliminate one parameter, the output was calculated for each option, not in terms of volume of water pumped, but in terms of hydraulic energy output; this effectively combines the volume of water pumped and the head, since units of $m^3.m$, or cubic metre-metres were used. However it should be realised that this is only valid for comparing similar systems; you cannot realistically compare systems operating at radically different heads such as a 100m borehole pump with a 10m head surface-suction pump purely on the basis of $m^3.m$. To convert a figure in $m^3.m$ to flow at a specific head it is only necessary to divide by the head in question; e.g. $10m^3.m$ could be $2m^3$ pumped through 5m head. To convert a unit cost of 5cents/$m^3.m$ to obtain a cost per unit of water, it is necessary to multiply by the head in question; eg. that energy cost at 2m head represents a water cost of 10c/m^3.

The final results are presented in terms of output cost versus the hydraulic energy demand. This is because the unit costs of different options, and hence the economic rankings, are sensitive to the size of system used. Therefore the choice of technology will differ depending on the scale of operation; systems which are economic for larger scale operations are often uneconomic on a small scale, and vice-versa.

Figs. 164 to 168 show the results for the different options analysed. In some cases, such as solar and wind powered pumping systems, the variability of the energy resource was allowed for by recalculating the band of results three times, i.e. for a mean of 10, 15 and 20MJ/m^2 per day (2.8, 4.2 and 5.6kWh.m^2 per day) of solar irradiation and, similarly for three mean wind-speeds of 2.5, 3.0 and 4.0m/s. The lower levels chosen are deliberately selected because they are sub-marginal conditions, while the middle level was judged to be marginal rather than attractive for the technologies concerned; so the results of all except the 20MJ/m^2 per day (for solar) and the 4.0m/s (for wind) examples would not be expected to show these technologies particularly favourably.

Table 41 COST AND PERFORMANCE ASSUMPTIONS USED FOR COMPARISON OF ALTERNATIVE PUMPING METHODS

Capital Cost	Life years	Maintenance per pump	Operating cost	Performance assumptions	Notes
Solar PV (hi) present Module $10/Wp Motor and pump $(500 + 1.5Wp) b.o.s. $(1500 + 2.0Wp)	15 7.5 15	$50+ $0.05/Wp p.a.	NIL	Motor/pump Subsystem efficiency = 35%	Sized for the design month. Irradiation levels of 20 MJ/m^2 in the design month examined. Peak water requirement in design month assumed to be 2 x average water requirement. Wp is the array rating in peak Watts.
Wind (hi) $400/m^2 of swept rotor area	20	$50 + 2.5 x area p.a.	NIL	Meaning Hydraulic power = 0.1 V^3 W/m^2 (= mean wind speed)	Sized for a design mean wind speed (in the design month) of 3 m/s. Peak water requirement = 2 x average water requirements.
Diesel (hi) $(1900 + 8.6P)	5	$400 p.a.	80 c/litres	Overall (hydraulic/ fuel) efficiency = 0.03 + 0.007P	P is the shaft power in kW. Minimum operation 0.25 hours per day. Efficiency in first half hour assumed to be half of 'steady-state' efficiency.
Diesel (lo) $(950 + 4.3P)	7.5	$200 p.a.	40 c/litre	efficiency = 0.13 + 0.007P	
Kerosene (hi) $600	2	$200 p.a.	80 c/litre	efficiency = 2%	Size of engine = 1.0 kW. Number of engines chosen to meet the demand.
Kerosene (lo) $200	5	$100 p.a.	40 c/litre	efficiency = 6%	Efficiency assumptions at start-up as for diesel.
Biogas Gas holder $(137 + G) b.o.s. $(91 + 1.89G) pump hi $600 pump lo $200	5	$20 p.a.	0.03 c/MJ of delivered gas	—	G is the energy content of the gas produced in MJ per day. The biogas unit is sized to provide the daily input energy requirements of the spark ignition engine.

Table 41 Continued

Capital Cost	Life years	Maintenance per pump	Operating cost	Performance assumptions	Notes
Oxen (lo) animal $250 pump $100	10 5	$20	$0.75 per animal day	Hydraulic output = 200W per animal	Assumed to work continuously for 8 hours per day at a hydraulic output of 200W i.e. 587m per day
Oxen (hi) animal $125 pump $100	10 5	$20	$1.25 per animal day		
Human (hi) $200 per pump	6		$1 per man day	Output of each pump = 37W hydraulic	Pump gives rated output for 4 hours per day (i.e. 54m per day)
Human (lo) $20 per pump	4	$20 p.a.	$0.30 per man day		
Turbine (lo) $200 per pump $200 for civil works	15 30	$20 p.a.	—	Output 350W per pump	These pumps are assumed to operate over a 24 hour period.
Turbine (hi) $200 per pump $2000 for civil works	15 30	$20 p.a.			
Hydram (hi) $3000 per hydram	30	$5 p.a.	—	Output 100W	
Hydram (lo) $1000 per hydram	30	$5 p.a.			
Mains (hi) $10000 for connection + $(265 + 0.75P) per pump	30	$20 p.a.	$0.042 per MJ of electricity = 15c/kWh		P is the rated output power of the pump. The pump size is determined by assuming the daily water requirements is to be provided in 6 hours.
Mains (lo) No connection charge $(265 + 0.75) per pump	30	$20 p.a.	$0.02 per MJ of electricity 7c/kWh		

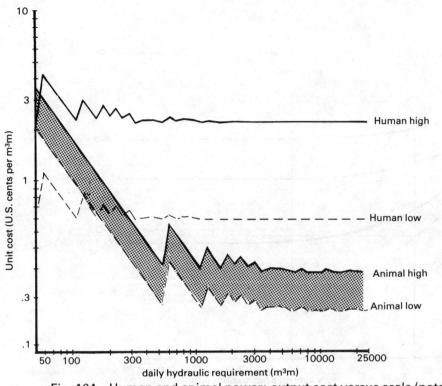

Fig. 164 Human and animal power: output cost versus scale (note log-log scale)

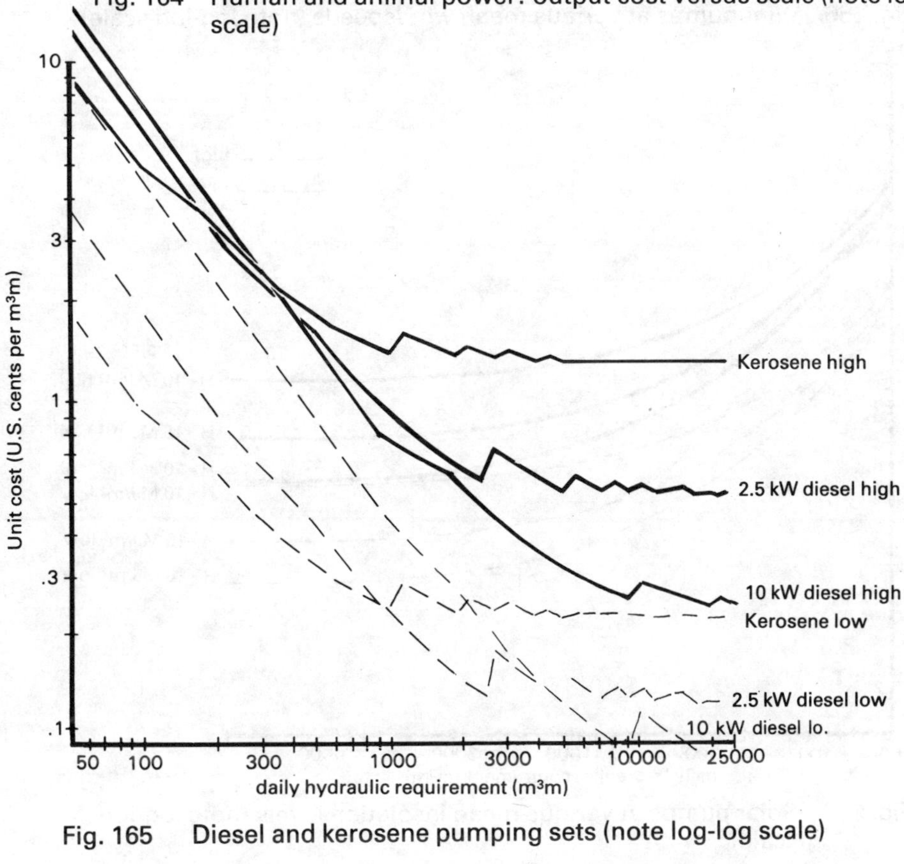

Fig. 165 Diesel and kerosene pumping sets (note log-log scale)

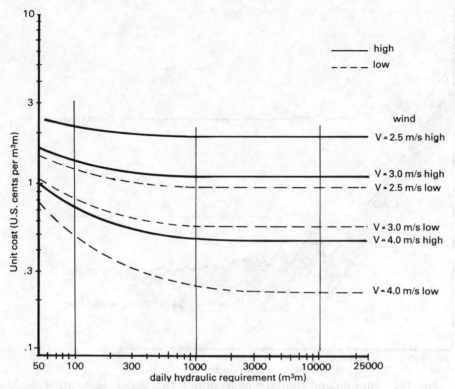

Fig. 166 Windpumps at various mean windspeeds (note log-log scale)

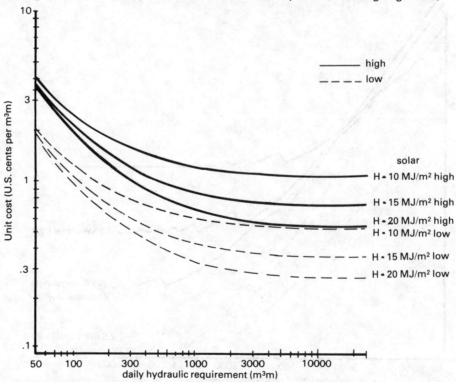

Fig. 167 Solar pumps at various mean insolation levels (note log-log scale)

A problem with Figs. 164 to 168 is that they had to be plotted on a log-log scale, because of the large range of power and cost considered, or either a very large sheet of paper would be needed to show the results, or the results at the lower end, which are of great interest to many people, would have been compressed to insignificance. The trouble with log-log scales is that the eye interprets distances linearly, so they can be misleading if simply inspected. This makes it difficult to compare the various options shown in Figs. 164 to 168. Therefore, Fig. 169 has been provided as a simplified composite of these results, using mean values (between the highs and lows) of the other graphs (to avoid too much of a confusion of curves) and moreover it was plotted against linear axes over a necessarily smaller size range, up to only 1,000m³.m/day. This range is of most interest as the relative rankings do not change much once an energy demand of about 1,000m³.m/day is exceeded.

Another, perhaps more easily interpreted presentation of these results is given in Fig. 170, where histograms of the cost spread for each system at daily energy demands equivalent to 100, 1,000 and 10,000 m³.m are given, and compared linearly rather than logarithmically for ease of comparison. (To put this in perspective, we are a considering say, 20, 200 and 2,000m³ per day at 5m head, or half those amounts at 10m head, etc.). This set of histograms also reintroduces the "optimistic" to "pessimistic" spread for each technology, which was lost in the previous comparison of Fig. 169. It is important not to lose sight of the possible range of costs applicable to any given technology, especially as in some cases the band of possible costs is very wide even on the basis of quite plausible assumptions in all cases.

It is interesting to see, for example, how a 10kW diesel system is by far the most expensive at the smallest demand level, where wind and solar pumps are at least competitive, but when 10,000 m³.m are needed the situation is completely reversed.

i. <u>Conclusions to be drawn from economic analysis</u>

The economics for most options are particularly size-sensitive, so that what is correct at 100m³.m/day is not generally true for a hydraulic requirement of 10,000m³.m/day.

The only low unit-output cost options which apply almost right across the entire size range of interest are:

a. mains electricity; providing it is already close to the field so only minor connection costs are incurred

b. hydro-powered devices (hydrams or turbine pumps); but depending on suitable site conditions

c. windpumps; but only for locations with high mean wind speeds (i.e. greater than 4m/s)

d. animal power is cost-competitive for energy demands exceeding about 500m³.m/day; but it does not generally seem a realistic new option where animal traction has not traditionally been used.

e. human power is cost-competitive in very small-scale applications (under 100m³.m/day); but only if a very low "opportunity cost" is assigned to human labour, and this conflicts with many development goals.

Where land-holdings are so small that the demand is less than 100m³.m/day, then human power is relatively inexpensive and animal power appears to be competitive. Solar and windpumps are both potentially competitive and so are spark ignition engines, providing they are reasonably efficiently sized and operated. Diesel power is not generally cost-effective for such small energy demands. Although the renewables in some cases appear competitive at this small demand level, the absolute costs of water are still rather high and it is important to ensure that irrigation will in fact produce a profitable yield in relation to the high water costs involved. It may be better to try to consolidate a single larger water system shared between several such small land holdings where such an option is feasible.

At the medium size range analyzed, namely 1,000m³.m/day, all the options are generally more cost-effective than they are at 100m³.m/day energy demand, and there is an overlap between most options; animal power, windpower (with V greater than 4m/s), water power, i.c. engines (only if efficiently operated) and mains electricity appear marginally the best options.

At the large size range of 10,000m³.m/day, diesel power comes into its own, and unless mains electricity or water power is available, diesel will probably be the best option.

Therefore, in summary, mains electricity (providing no connection costs are involved) or water power are most economical. Windpower is next most attractive if windspeeds are high, (but it is decidedly unattractive with low or uncertain winds). Solar power is generally expensive but has the potential to fill a useful gap in the 100-1000m³.m/day demand level range once the cost of solar-powered systems falls a little more. Engines have a very wide band of uncertainty relating to their unit costs at small energy demand levels, ranging from competitive to unacceptable. Spark ignition engines are more attractive in the small to medium range of 100-1,000m³.m/day while diesel engines become more competitive at energy demands exceeding around 1,000m³.m/day. Biomass-fuelled spark ignition engines will generally cost more to run than kerosene or gasoline fuelled engines (where fuel is at world prices), but will be worth considering where petroleum fuels are either not available or have a high opportunity cost; obviously a suitable low-cost biomass fuel resource needs to be readily available.

5.2 PRACTICAL CONSIDERATIONS

It is worth elaborating on some of the practical, in addition to economic, considerations that relate to the different types of water lifting system.

5.2.1 Status or Availability of the Technology

Some technologies are more "available" than others. Table 5.4 indicates technologies in general use, or with "future potential", plus some

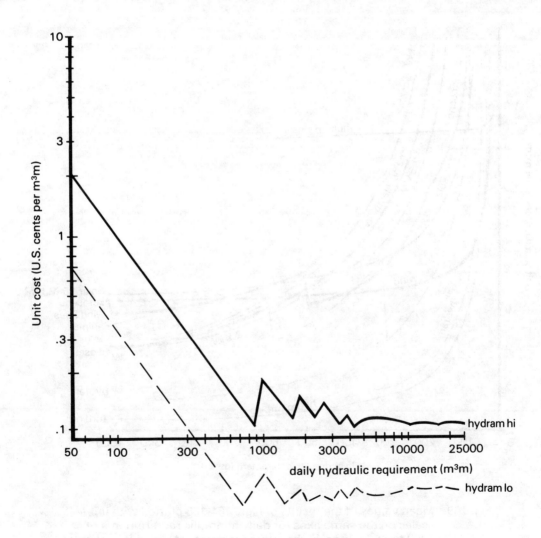

Fig. 168   Hydrams (note log-log scale)

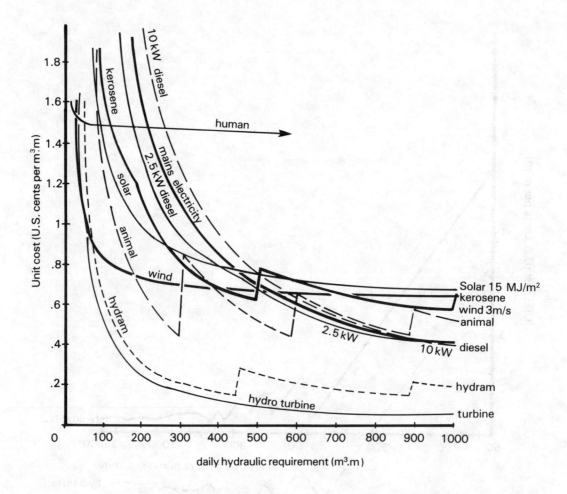

Fig. 169 Mean values of the results of Figs. 164-168 plotted with linear scales on the same axes for daily hydraulic requirements of 1000 m³m. 'Jumps' in the curve are points at which it becomes necessary to use another unit to meet the demand.

that are obsolete; a few qualify for more than one of these categories. In some cases, such as small stirling engines, it is believed that not one commercial manufacturer currently offers a viable product even though in this case they were widely used in the past and there seems no techno-economic reason why they would not be attractive today. Similarly, the Chinese Turbine Pump, which is widely used in southern China, appears to be most attractive economically (and has numerous operational advantages) but is not currently produced outside China. Therefore, for some time to come, such technologies may only generally be considered as a potential option in most countries.

A more general problem with all new or unfamiliar technologies, even if they are notionally commercially available, is to obtain the necessary information and advice in order to:

a.   procure the correct size and specification of system

b.   install it correctly

c.   operate and maintain it effectively

It is probably best if all but the more adventurous (and wealthy) of small farmers play safe and stick to familiar and "available" technologies where help, advice and spares are readily available and risks are minimized. However, if everyone took this advice, new and perhaps eventually better technologies would never become available, so therefore it is worth suggesting that it is necessary for government, international aid agencies and institutions, with a commitment to the future development of small-scale agriculture, to take risks in this area on behalf of their local farmers and to test and demonstrate any technologies that appear promising in the local irrigation context.

Problems must be expected with pilot projects. It is therefore vital to measure, monitor and record the behaviour of any innovative systems that are tried. Even if no problems occur, unless such pilot projects are properly monitored it will not be possible to come to any conclusion as to whether the new technology being tried is competitive with what it is supposed to replace.

Actual performance monitoring is important, but so are qualitative comments on operational aspects, such as maintenance or installation difficulties, or shortcomings as perceived by the user. Feedback on these aspects needs to be absorbed by the manufacturers and developers of this kind of equipment, so that the necessary improvements can be set in hand, otherwise development will be delayed.

5.2.2. Capital Cost versus Recurrent Costs

As explained previously, low recurrent costs tend to have to be traded for high capital costs. High capital costs represent a real barrier for small farmers to take up new technology even if the unit output costs are competitive. Worse, low capital costs are often an incentive to install inefficient systems (eg. small kerosene pumps sets with inadequate distribution pipe). Where there is a good case for farmers to be encouraged to use a high capital cost technology (even to move from spark-ignition to diesel engines), then it will generally be necessary for appropriate credit facilities to be made available as an incentive.

Renewable energy systems, with their high capital costs and low recurrent costs may be of particular interest to institutions having access to grant or aid finance for capital items, because they do offer a means for investing in low recurrent costs. Many rural institutions face major problems with meeting the recurrent costs of running conventional pumping systems, so in some situations it may make sense to introduce high capital cost equipment simply to reduce the maintenance and fuel budgets.

5.2.3 Operational Convenience

This factor varies considerably with different types of pumping system. For example, a windpump will be highly dependent on adequate wind to operate, so if high risk crops are grown where the provision of water on demand is vital to the survival of the crop, then a large (and consequently expensive) storage tank will be necessary to ensure water is always available. Alternatively less risky and probably less valuable crops could be grown with only a small storage tank or even in some cases with no storage at all. Therefore the flexibility of the device, or its controlability, must be taken account of as they affect such fundamental decisions as the choice of crop to grow under irrigation.

Other factors relate to aspects such as size and portability. Small engines and small solar pumps may be quite portable, which means they can be moved around to irrigate with only short, but effective distribution pipes, while a windpump, a larger engine, or a hydram will inevitably have to be fixed. However small portable items are also vulnerable to theft in some regions, which makes the relatively large and fixed installation less at risk in that context.

Few options can rival the operational flexibility of an i.c. engine powered system in terms of rapid start up, portability, provision of power on demand, etc., but of course one of the reasons for looking at the other options is that the i.c. engine generally suffers the major drawback of needing petroleum fuels. So the operational shortcomings of many of the alternatives need to be weighed against the fuel needs and the likely future availability and cost of fuel.

5.2.4 Skill Requirements for Installation, Operation and Maintenance

Two key factors apply here; the absolute skills required and the level of familiarity with the equipment. Commonplace, but complex machines like diesel engines can often appear to be simpler to handle than much simpler but less familiar technologies such as solar pumps (from the maintenance point of view). Due allowance for the need to learn about a new technology must therefore be made before dismissing it as too sophisticated. In absolute technical terms there are no water lifting technologies more technically demanding than the diesel engine; most of the renewables are basically much simpler even if in a few cases they involve little understood concepts (not many diesel mechanics understand the first principles of a diesel engine either - it is not necessary to know this in order to overhaul an engine).

The level of support available from manufacturers or suppliers is most important; most successful technologies have become widely used because they

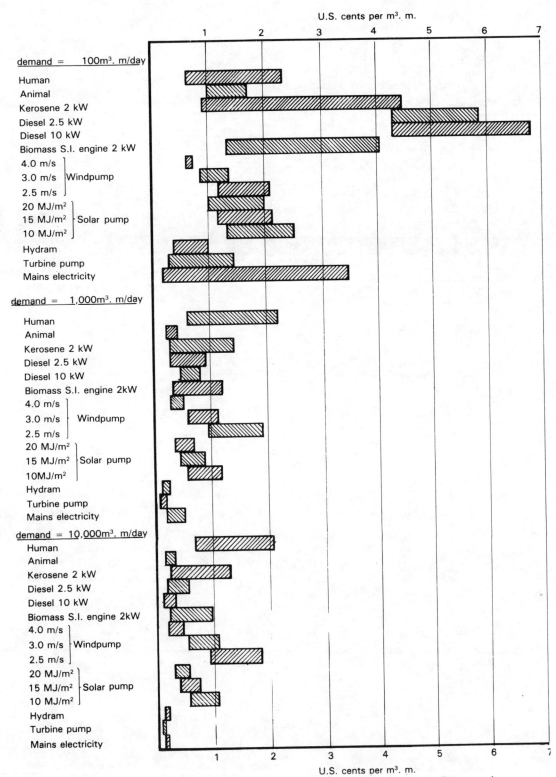

Fig. 170 Expected range of unit energy costs for three levels of demand, if 100, 1000 and 10,000 m³m/day for different types of prime-mover

were successfully promoted and supported by the commercial interests that market them. Even the simplest technologies stand little chance of being taken up unless they are effectively promoted and the early users are also properly supported and helped through any problems.

Often there are trade-offs between the amount of skill needed and the amount of maintenance/operational intervention required. For example; solar pumps need very little maintenance, but what they do need is usually unfamiliar and demands specialized and at present rare skills, (although given appropriate training the necessary skills can be rapidly attained); i.c. engine pumps need quite frequent and sometimes technically sophisticated maintenance functions, but because the technology is so widespread there are many people capable of performing these; the cruder types of village built windpumps need a great deal of adjustment and running repairs, but to the people who are familiar with them, these present little inconvenience or difficulty.

Familiarity or lack of familiarity are perhaps more important factors then the absolute level of technical skill needed, (after all few repairs can be more demanding then overhauling a diesel injection pump or reboring an i.c. engine, yet few provincial towns exist in developing countries where such activities cannot be carried out). Therefore training is a vital aspect of introducing any new or unfamiliar technologies.

5.2.5 Durability, Reliability and Useful Life

Durability, reliability and a long operational life usually cost money, but they also are frequently a good investment in terms of minimizing costs. Perceptions on the value of capital or the choice of discount rate will usually control decisions on these aspects. However it is best to try and show through economic or financial analysis whether the benefits from buying high quality equipment are cost-effective. Many who have analyzed the cost of operating machinery in remote areas, particularly engines, have concluded that their performance, not only in terms of output but also in terms of reliability and durability, usually turns out to be sigificantly worse than expected. There is therefore often merit in erring towards oversizing prime-movers and procuring any special accessories that make the system more "fail-safe".

5.2.6 Potential for Local Manufacture

This is of more immediate interest to policy makers than farmers, although the benefits to be gained from local manufacture could of course eventually benefit the latter.

One of the principal reasons to seek alternatives to petroleum fuelled engines for irrigation is because of the inability of many countries to import sufficient petroleum to meet present, let alone future, needs. Therefore shortage of foreign exchange to import oil equally implies shortage of foreign exchange to import foreign solar pumps or other such alternatives. Therefore, any system which lends itself to whole or partial local manufacture is of potential economic importance in terms of import substitution for oil, (and probably for engines too). However, the benefits of local manufacture do not end with import substitution.

Other important results of local manufacture, or part-manufacture, are:

- creation of local industrial employment
- enhancement of industrial skills
- improved local availability of spare parts
- improved local expertize in the technology

in other words, local manufacture can help to overcome many of the constraints mentioned previously in supporting the initial diffusion of a new technology, and can at the same time help to develop the industrial/manufacturing base. The economy of the country gains twice from local manufacture of irrigation equipment, first from internalising the manufacture and secondly from the enhanced agricultural production once the equipment starts to be widely applied.

5.3 CONCLUSION

The reader will have realized from this paper that there is a wide variety of options for combining prime movers and water lifting devices in order to pump irrigation water. In practical terms the situation will be simplified for most people by having to choose from a much more limited selection of what is available rather than what is possible. However, it is hoped that by applying an understanding of the technical and economic principles described, more cost-effective irrigation may be achieved, and hence more irrigation.

Small scale lift irrigation is not yet normally practiced in many countries, so the incentive to develop the full range of potentially useful technologies has not yet been given to industry. It is to be hoped that the increasing need to grow more food will drive irrigation technology forward and result during the next decade or two in a considerable widening of the choice of equipment on the market. The author hopes this book will have made a small contribution to encouraging the understanding and interest needed for this to happen.

REFERENCES

1. Molenaar, A., Water Lifting Devices for Irrigation, FAO, Rome.
   1956
2. World Bank, Energy in the Developing Countries, Washington, DC.
   1983
3. Stern, P S, Small Scale Irrigation, I T Publications, London.
   1979
4. World Bank, World Development Report - 1982, World Bank/OUP, Oxford.
   1983
5. World Bank, World Development Report - 1980, World Bank/OUP, Oxford.
   1981
6. Leech, G., Energy & Food Production, IIED, London.
   1975
7. World Water, The Right to Rice, Liverpool, Oct 1980.
   1980
8. Booher, L.J., Surface Irrigation, Agricultural Development Paper No. 95,
   1974   FAO, Rome.
9. Kraatz, D.B., Irrigation Canal Lining, Land and Water Development Series
   1977   No. 1, FAO, Rome.
10. Vermeiren & Jobling, Localized Irrigation, Irrigation & Drainage Paper
    1980   No. 40, FAO, Rome.
11. Doorenbos & Pruitt, Crop Water Requirements, Irrigation & Drainage Paper
    1977   No. 24, FAO, Rome.
12. Kraatz, D.B., Socio-Economic Aspects - cost comparison and selection of
    1981   water lifting devices, Proc. FAO/DANIDA Workshop on Water
              Lifting Devices in Asia and the Near East, held in Bangkok Dec
              1979, FAO, Rome.
13. Roberts, W. & Singh, S., A Text Book of Punjab Agriculture, Civil &
    1951   Military Gazette, Lahore, Pakistan.
14. Mead, Daniel W., Hydraulic Machinery, McGraw-Hill, New York and London.
    1933
15. Fraenkel, P.L., Food from Windmills, IT Publications, London.
    1976
16. Collett, J., Hydro Powered Water Lifting Devices for Irrigation, Proc.
    1981   FAO/DANIDA Workshop on Water Lifting Devices in Asia and the
              Near East, held in Bangkok Dec 1979, FAO, Rome.
17. Watt, S.B., Chinese Chain & Washer Pumps., I T Publications, London.
    1977

18. Gao Shoufan, Yin Jianguo, and Cheng Zuxun, Tube-chain waterwheel, proc.
1982   UNDP/FAO China workshop of Nov, FAO, Rome.

19. Dunn, P.D., Appropriate Technology, Macmillan, London.
1978

20. Intermediate Technology Development Group, Water for Rural Communities,
1983   Appropriate Technology Journal, Vol 9, No. 1, IT Publications, London.

21. Hofkes, Manual Pumping of Water for Community Water supply and small
1981   scale irrigation, proc. FAO/DANIDA Workshop on Water Lifting Devices in Asia and the Near East, held in Bangkok Dec 1979, FAO, Rome.

22. Morgan, letter in Appropriate Technology Journal Vol 9 No 1, IT
1983   Publications, London.

23. Wilson, S.S., Pedalling foot power for pumps, World Water, Liverpool,
1983   Nov 1983.

24. Schioler, T., private communication.
1985

25. Khan, H.R., Study of Manual Irrigation Devices in Bangladesh, Proc. AT in
1980   Civil Engineering conf., Inst. of Civil Engineers, London.

26. McJunkin, Handpumps for use in Drinking Water Supplies in Developing
1977   Countries, IRC, The Hague.

27. UNDP/World Bank, Global Handpump Project, Technical Paper No. 3.,
1982   Washington DC..

28. Klassen, The Rower Pump, proc. FAO/DANIDA Workshop on Water Lifting
1981   Devices in Asia and the Near East, held in Bangkok Dec 1979, FAO, Rome.

29. Tata Energy Research Inst., Energy Update, Bombay.
1979

30. Birch D.R. & Rydzewski J.R., Energy Options for Low Lift Irrigation Pumps
1980   in Developing Countries: The case of Bangladesh & Egypt, ILO, Geneva.

31. Hood, O.P., Certain Pumps and Water Lifts used in Irrigation, U S
1898   Geological Survey, Washington DC.

32. Jansen, W.A.M., Performance Tests of Kerosene Pumpsets Wind Energy Unit,
1979   Colombo, Sri Lanka.

33. Reader, G.T., Hooper C., Stirling Engines, E & F N Spon, London and New
1983   York.

34. Ross, A., Stirling Cycle Engines, Solar Engines, Phoenix, Az.
1977

35. Fluitman, F., The Socio-Economic Impact of Rural Electrification in
    1983   Developing Countries: A Review of the Evidence, ILO, Geneva.

36. Vadot, L., Water Pumping by Windmills, (translation of 'Le pompage de
    1957   l'eau par eoliennes'), La Houille Blanche, No.4, pp 496-535,
             Grenoble, France, Sept.

37. Perry, Experiments with Windmills, US Geological Survey, Washington, DC.
    1899

38. Sidney Williams & Co, Comet Catalogue No. 9, Dulwich Hill, NSW,
    1968   Australia.

39. Eldridge, Frank R., Wind Machines, van Nostrand Reinhold, New York.
    1980

40. Fraenkel, P.L., Wind Technology Assessment Study, I T Power Ltd.,
    1983   Reading for UNDP/World Bank, Washington, DC..

41. Gao Rushan, Zhu Zhongde, On the Development of Human, Animal, Wind and
    1979   Water Power Lifting Devices for Irrigation and Drainage in
             China, Ministry of Water Conservancy, Beijing.

42. Schioler, T., Water Lifting Devices for Irrigation - History and General
    1981   Background, proc. FAO/DANIDA Workshop on Water Lifting Devices
             in Asia and the Near East, held in Bangkok Dec 1979, FAO,
             Rome.

43. Gilmore, E., Barieau, R. E., and Nelson, V., Final Report on the
    1977   Feasibility of using Wind Power to Pump Irrigation Water, for
             the Governor's (of the State of Texas) Energy Advisory
             Council, Austin, Texas.

44. Nelson, V., et al., Wind Power Applications in the US - Irrigation
    1982   Pumping, Wind Engineering, Vol 6 No2.

45. Lysen, E., Introduction to Wind Energy, CWD, (formerly SWD), PO Box 85,
    1983   Amersfoort, The Netherlands.

46. Pinilla, et al., Wind Energy and Water Pumped: Conversion Efficiency
    1984   Limits using Single Acting Lift Pumps, Proc. Conf. BWEA,
             Cambridge University Press, Cambridge.

47. Golding E.W., The Generation of Electricity by Windpower, E & F N Spon,
    1976   London (revised edition).

48. World Meteorological Office, Technical Note No. 175, Meteorological
    1981   Aspects of the Utilization of Wind as an Energy Source,
             Secretariat of the WMO, Geneva.

49. Leicester, R. J., Study of the World Market for Medium-sized Wind
    1981   Generators, for Overseas Development Administration, Crown
             Agents, London.

50. Lipman, N.H., et al, (Editors), British Wind Energy Association, Wind
    1982   Energy for the Eighties, Peter Perigrinus, Stevenage, UK and
             New York.

51. ECDC-TCDC, Renewable Sources of Energy, Volume III, Wind Energy, UN
1981   Economic and Social Commission for Asia and the Pacific, Bangkok.

52. Butti, K,. & Perlin, J., A Golden Thread: 2500 Years of Solar Energy and
1981   Technology Marion Boyars Pubs., London.

53. Daniels, F., Direct Use of the Sun's Energy, Yale University Press, and
1974   Ballantine Books, New York.

54. Halcrow/I.T.Power, Small Scale Solar Powered Irrigation Pumping Systems:
1981   Technical & Economic Review, World Bank, Washington DC.

55. Halcrow/I.T.Power, Small Scale Solar Powered Irrigation Pumping Systems:
1981   Phase I Project Report, World Bank, Washington DC.

56. Halcrow/I.T.Power, Small Scale Solar Powered Pumping Systems: The
1983   Technology, its Economics and Advancement - Main Report, World Bank, Washington DC.

57. Kenna, J.P. and Gillett, W.B., I.T.Power/Halcrow, Handbook on Solar
1985   Water Pumping, World Bank, Washington DC and I T Publications, London.

58. Rauschenbach, H.S., Solar Cell Array Design Handbook, van Nostrand
1980   Reinhold, New York.

59. World Meteorological Office, Meteorological Aspects of using Solar
1981   Radiation as an Energy Source, Technical Note 172, WMO Secretariat, Geneva.

60. Rosenblum, Practical Aspects of Photovoltaic Technology, Applications,
1985   and Cost, NASA Lewis and USAID Science and Technology Bureau Office of Energy, Washington DC.

61. Meier, U., Local Experience with Micro-hydro Technology SKAT, St Gall.
1981

62. Alward, et al., Micro-hydro Power: Reviewing an old Concept, NCAT/US
1979   Dept. of Energy, Butte, Montana.

63. Summary Report, Proceedings of the FOA/UNDP/China Workshop on Water
1982   Lifting Devices and Water Management, Held in China Nov 1981, FAO Rome.

64. Tacke, J.H.P.M., Hydraulic Rams, Delft University of Technology, draft
1984   for IRC publication Alternative Energy Resources for Rural Water Supply Applications in Developing Countries, The Hague.

65. Nguyen Trong Lac, Norias in Vietnam, proc. FAO/DANIDA Workshop on Water
    1981   Lifting Devices in Asia and the Near East, held in Bangkok Dec 1979, FAO, Rome.

66. Collett, J., Water Powered Water Lifting Devices for Irrigation,
    1980   unpublished revision for FAO of ref. [16], Intermediate Technology Consultants, London.

67. Fraenkel, P., & Musgrove, P, River & Tidal Current Turbines, Proc. Int.
    1979   Conf. on Future Energy Concepts., Inst. of Electrical Engineers, London.

68. Hall, D.O., World Biomass: An Overview, Proc. Biomass for Energy, UK
    1979   International Solar Energy Society, London.

X 69. Kristoferson, L., Bokalders, V., Newham, M., Renewable Energy for
    1984   Developing Countries: A Review, The Beijer Inst., Stockholm.

70. FAO, Map of Fuelwood Situation in Developing Countries, FAO, Rome.
    1981

71. Brown, Lester R., Food or Fuel, Paper No.35, Worldwatch Institute,
    1980   Washington DC..

72. Earl, D.E., Forest Energy & Economic Development, Oxford University
    1975   Press, Oxford.

73. Ascough, W.J., Producer Gas for Automotive Use, Proc. Alternative Fuels
    1981   for I.C.Engines, Reading University and ITIS, Reading and Rugby, England.

74. Foley, G., Barnard, G., and Timberlake, Gasifiers: Fuel for Seige
    1983   Economies, Earthscan/IIED, London.

75. van Swaaij, W.P.M., Gasification: the Process and the Technology, Proc.
    1980   Energy for Biomass, Brighton.

76. Damour, V & M-S, Development of Small Scale Gasifiers for Irrigation in
    1984   India, Paris.

77. Coombs, J., Production of Alcohol Fuels, Proc. Alternative Fuels for
    1981   I.C.Engines, Reading Univ. and ITIS, Reading and Rugby.

78. Radley, R.W., et al, Vegetable Oils as Fuels for Diesel Engines,
    1981   Proc.Alternative Fuels for I.C.Engines, Reading University and ITIS, Reading and Rugby.

79. Hall, D.O., Vegetable Oils for Diesel Engines, Private Paper, Kings
    1981   College, London University.

80. Goddard, K., Liquid Fuels from Biomass, Chartered Mechanical Engineer,
    1979   London, Jan 1979.

81. Hall, D.O., Renewable Resources (Hydrocarbons), Outlook on Agriculture,
    1980   Vol 10, No. 5.

82. Pyle, D.L., Biogas & Thermochemical Products, Proc. Alternative Fuels
    1981   for I.C.Engines, Reading University and ITIS, Reading and Rugby.

83. van Buren, A., A Chinese Biogas Manual, I T Publications, London.
    1979

84. Meynell, P-J., Methane: Planning a Digester, Prism Press, Dorchester.
    1982

85. UN ESCAP, Renewable Sources of Energy: Volume II, Biogas Bangkok.
    1981

86. ITDG Mission to China, Internal Report, Intermediate Technology
    1981   Development Group, London.

87. Hofkes, E.N., et al., "Renewable Energy Sources for Rural Water Supply
    1986   in Developing Countries", International Reference Centre for
            Community Water Supply and Sanitation, the Hague.